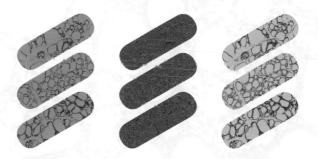

OXFORD BIOLOGY PRIMERS

Discover more in the series at
www.oxfordtextbooks.co.uk/obp

Published in partnership with the Royal Society of Biology

Royal Society of Biology

MICROBIAL BIOTECHNOLOGY

OXFORD BIOLOGY PRIMERS

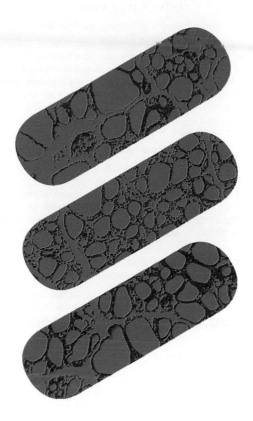

MICROBIAL BIOTECHNOLOGY

Kay Yeoman, Beatrix Fahnert, David Lea-Smith, and Tom Clarke

OXFORD
UNIVERSITY PRESS

Royal Society of
Biology

OXFORD
UNIVERSITY PRESS

Great Clarendon Street, Oxford, OX2 6DP,
United Kingdom

Oxford University Press is a department of the University of Oxford. It furthers the University's objective of excellence in research, scholarship, and education by publishing worldwide. Oxford is a registered trade mark of Oxford University Press in the UK and in certain other countries

Impression: 1

Published in the United States of America by Oxford University Press
198 Madison Avenue, New York, NY 10016, United States of America

British Library Cataloguing in Publication Data
Data available

Library of Congress Control Number: 2020945535

ISBN 978-0-19-882281-3

Printed in Great Britain by
Bell & Bain Ltd., Glasgow

SERIES PREFACE

Welcome to the Oxford Biology Primers

There has never been a more exciting time to be a biologist. Not only do we understand more about the biological world than ever before, but we're using that understanding in ever-more creative and valuable ways.

Our understanding of the way our genes work is being used to explore new ways to treat disease; our understanding of ecosystems is being used to explore more effective ways to protect the diversity of life on Earth; our understanding of plant science is being used to explore more sustainable ways to feed a growing human population.

The repeated use of the word 'explore' here is no accident. The study of biology is, at heart, an exploration. We have written the Oxford Biology Primers to encourage you to explore biology for yourself—to find out more about what scientists at the cutting edge of the subject are researching, and the biological problems they're trying to solve.

Throughout the series, we use a range of features to help you see topics from different perspectives.

Scientific approach panels help you understand a little more about 'how we know what we know'—that is, the research that has been carried out to reveal our current understanding of the science described in the text, and the methods and approaches scientists have used when carrying out that research.

Case studies explore how a particular concept is relevant to our everyday life, or provide an intimate picture of one aspect of the science described.

The bigger picture panels help you think about some of the issues and challenges associated with the topic under discussion—for example, ethical considerations, or wider impacts on society.

More than anything, however, we hope this series will reveal to you, its readers, that biology is awe-inspiring, both in its variety and its intricacy, and will drive you forward to explore the subject further for yourself.

PREFACE

It is fascinating how biotechnology and its applications are fundamental to our way of life. Ancient biotechnology processes such as the production of cheese and cured meats have been vital in food preservation. The production of beer was important in providing a disease-free source of drinking water, so humans could flourish, culturally develop, and achieve. Whilst these are ancient processes, our understanding of them has developed as we have learnt about microbes and their roles in the environment. Gradually we started to manipulate the processes to get better, more reliable products. In the early 20th century our understanding had reached a level where the first deliberate biotechnology processes could be designed—the acetone-butanol fermentation, developed using *Clostridium acetobutylicium*, quickly followed by citric acid production using *Aspergillus niger* in the 1920s. The last 100 years has seen an explosion of products produced by microbes on an industrial scale, from pharmaceuticals to fine chemicals such as enzymes. And the exploration and application continues. Microbes will play a vital role in the future of agriculture and the production of edible food through single-cell protein.

This book will take you though a biotechnology journey. You will start in Chapter 1 by getting to grips with the fundamentals of designing a biotechnology process, the hosts, the media, and the types of reactor and their operative conditions. You will start to appreciate the constraints imposed by moving from laboratory-scale to industrial-scale production. Microbial cells have to be analysed, and it is important to know when during growth the desired product will be produced. Chapter 2 will take you through how microbial cells are monitored and the basics of microbial growth kinetics which allows the manipulation of specific growth phases to increase product yield. Chapters 3 and 4 will increase your depth of knowledge around specific biotechnology processes to produce a whole range of biologics, fine chemicals, and components for food. We will examine the future of microbes as a source of protein. Chapter 5 focuses on the role that biotechnology can play in the environment, and you will learn about the importance of wastewater treatment, and how microbes can be used to clean up damaged environments from, e.g., oil spills. Chapter 6 will introduce you to the ideas around synthetic biology and how the host organism can be manipulated to produce more of a desired product, or indeed how a product such as spiders' silk can be engineered in a microbe. Chapter 7 takes you into the world of diagnostics, and how microbes are being used to produce molecules such as antibodies and enzymes which are used in diagnostic tools, for example to detect the Ebola virus. Chapter 8 will introduce you to concepts around the usefulness of microbes in agriculture, from microbial inoculants in the soil to promote nitrogen fixation to the use of microbes as biological control agents to control crop diseases. Chapter 9 will deepen your knowledge of how extremophiles are used in biotechnology. You are probably familiar with the *Taq* DNA polymerase enzyme used in PCR, but this chapter will take you further and will introduce you to cold-active enzymes, and those organisms which can be used for biomining of metals, which require extremely low pH for survival. Chapter 10 has been written to encourage your thinking around how microbes are constantly eroding our built environment, from our art to our building infrastructure, and what we can do to combat this destruction. Finally, Chapter 11 will enable you to consider the ethics of developing biotechnology applications, the risks posed to the environment, as well as issues around patenting.

You will see how biotechnology draws on many other subjects such as biochemistry, microbial physiology, molecular biology, and genetics. We will touch on these throughout the primer. To explore these subjects in more detail, and to consolidate your basic knowledge in these key areas, we recommend use of the appropriate primer in the series.

ABOUT THE AUTHORS

Professor Kay Yeoman teaches microbiology in the School of Biological Sciences at UEA. She has undertaken research in the use of agricultural waste materials in fermentation media. She has also conducted research in the Rhizobium:legume symbiosis, investigating the uptake of iron. She is keen communicator of science, leads fungal forays and is particularly interested in fungi and their use in industry.

Dr David Lea-Smith is a lecturer in microbiology at the University of East Anglia. His research is primarily focused on investigating photosynthetic cyanobacteria, specifically understanding their biochemistry and physiology, impact on the environment and potential for biotechnology as a platform for renewable production of industrial and medical compounds.

Dr Tom Clarke graduated from UCL in Biochemistry and undertook a PhD at the John Innes centre in Norwich. After a research fellowship at the University of Michigan he joined UEA biological sciences in 2003. His current research involves uncovering how bacteria generate electricity in different environments.

Dr Beatrix Fahnert is a microbiologist with a background in medical, industrial and applied microbiology. She has been teaching related subjects at Undergraduate and Postgraduate level for more than 20 years, and is currently an Associate Professor at the School of Life Sciences, University of Lincoln. Beatrix is a member of the European Federation of Biotechnology and the UK Microbiology Society, where she served as Chair-Elect of the Education Division.

CONTENTS

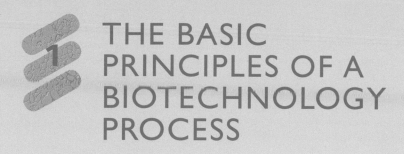

1 THE BASIC PRINCIPLES OF A BIOTECHNOLOGY PROCESS

Learning Objectives

- To explain the concept of bioprocess development with reference to the necessary strategic follow-through from upstream to downstream processing;
- to evaluate the choice of microbial host for the biotechnology process;
- to explain how microbial cells can be improved and manipulated by using genetic techniques;
- to evaluate fermentation media and fermenter design;
- to explain how biotechnology processes have to be scaled and optimized for industrial production of the desired product;
- to evaluate the recovery of the product through downstream processing.

When we think about a biotechnology process, very likely the image of a shiny steel reactor holding a volume of thousands of litres comes to mind. Interestingly, while this is a true example, it is only one way in which we can grow microbes and we will cover many more types of reactor and biotechnology processes in this chapter. The design of a biotechnology process is complex and many variables have to be considered. Throughout this book you will also be able to explore the basic principles, contexts, and relevance of biotechnology to business and society.

1.1 Overview of the biotechnology process

A **reactor** of any kind is only one part of the entire biotechnology process, which starts upstream with the identification of a promising target, and is continued via downstream processes until the desired product or outcome has been achieved. In Figure 1.1 you can see an overview of the bioprocess development. Every step requires input from teams of microbiologists to engineers.

Figure 1.2 gives some examples of how **productivity** is optimized during upstream development when choosing/creating a suitable **host species** and if necessary a **recombinant vector**, as well as designing the cultivation regime. We discuss these steps now, and then cover the downstream part at the end of the chapter.

Figure 1.1 Bioprocess development comprises of all upstream steps before cultivation and downstream steps after cultivation. More detail is provided in Figures 1.2 and 1.14.

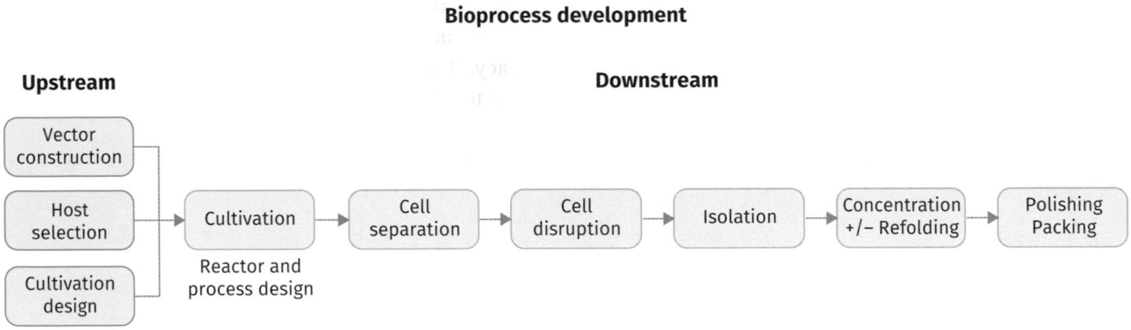

Figure 1.2 Upstream development steps. Vector construction is only necessary for recombinant hosts.

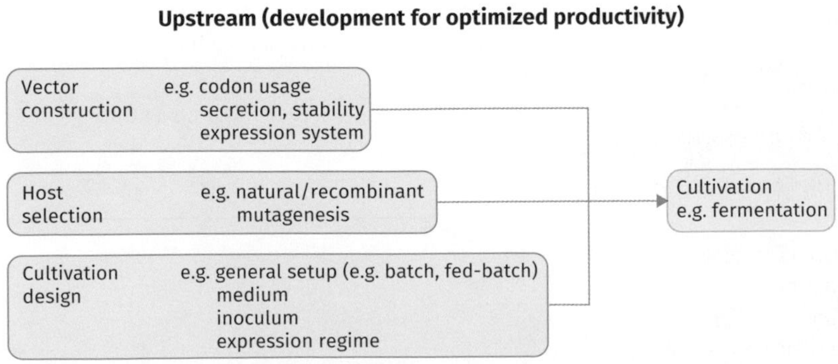

Any biotechnology process development is a compromise that allows delivery of the desired product or outcome by means of the most efficient cultivation as quickly as possible and at the lowest cost. The costs can be very high, due to process demands in the context of gaining approval, and if the necessary product integrity cannot be achieved otherwise. For **heterologous proteins** we want a high yield and very high purity, and authentic **post-translational modifications** might be necessary for biological activity, solubility, and stability. Such high-level requirements are often related to therapeutic proteins. By contrast, for proteins such as enzymes used in laundry the production process is much cheaper. The difference in cost may arise if the product is secreted into the culture medium instead of remaining in the host cell. Secretion may involve introducing a host system-related **signal sequence** into the vector, or replacing the usual signal sequence with that of a different **cell compartment**. To enhance or prevent solubility of the protein, its folding speed can be adjusted based on the **secretion system** or the **codon usage**. These are only a few illustrating examples. There are many factors and considerations that impact development of a successful biotechnology process. Traditionally the approach was trial and error, based on expert knowledge. For some time now Design of Experiment (DoE) has been used, which involves a statistical approach, often combined with High Throughput Methods (HTP) that allow testing a large number of alternatives for a certain variable at the same time. A concerted effort has to be made in development of upstream and downstream processing, because they are inevitably linked. One sub-optimal downstream choice may affect how the product needs to be purified and could dramatically increase the costs. That is why the Quality by Design (QbD) principle of manufacturing is applied, where the overall focus is to eliminate manufacturing risks.

This approach is particularly relevant if we think about **Good Manufacturing Practice** (GMP), which refers to minimum standards for everything related to the production process of pharmaceutical products, in order to ensure consistent high quality. This is a prerequisite for the approval of the product for sale, to ensure that it is not a health hazard and has clinical efficacy. It means the product does not contain any impurities or contaminations, or too low or too high activity. Written standard operating procedures for every step of the production process need to be in place, as well as evidence that those procedures were consistently followed. This starts with using certified medium ingredients from known sources, well characterized and stored organisms, sterilization efficacy, appropriate storage of components at all times, accurate labelling via documented process integrity, to staff training, including personal hygiene. The necessary compliance is enforced by local and broader agencies such as the World Health Organization, the European Medicines Agency, and the US Food and Drug Administration (FDA).

The bigger picture panel 1.1
Putting a price on health

Bringing a successful new pharmaceutical product to market takes years, can cost more than one billion USD, and only a small fraction of initially promising compounds are successfully commercialized. Companies take a considerable financial risk, which is then mitigated by long-term **patent** protection. The price of the sold drugs can be quite high because the initial costs have to be recovered over the life of the patent. This means that patients living in some developing countries or who have low economic status might not be able to afford treatment. That is why some companies produce drugs infringing specific patents and then sell the drugs at low cost. This affects the market and potentially limits the ability or willingness of companies to invest in developing new drugs. Moreover, some companies do not follow good manufacturing practice or may even produce counterfeit drugs, both of which pose a health risk. In the case of antibiotic production there could be a low amount of the active product which can lead to bacterial resistance.

What are the issues in this complex situation, and are there any that have not yet been alluded to? What are the challenges addressing any of these issues?

Key points

- A biotechnology process has an upstream and a downstream part.
- Every step of a bioprocess has to be optimized towards an overall, frequently quality controlled, compromise.
- Statistical approach to experimental design has started to replace trial and error.

1.2 The hosts

Microbial hosts are extremely useful in biotechnology processes for a number of reasons.

1. They are relatively easy to cultivate and when the right conditions are provided they can grow rapidly.
2. Space requirement is small. Large amounts of microbial biomass can be cultivated on limited land areas.
3. Many microbes can use cheap waste substrates on which to grow, e.g. bagasse, which is a waste product form the sugar production industry.
4. They can adapt to different environments.
5. They provide a great variety of different biochemical reactions.
6. There is a huge diversity of potential microbial products that are of use to humankind.
7. They can make specific isomers (usually active ones).
8. Many microbes can readily undergo genetic manipulation to improve strains.
9. Their high volume to surface ratio facilitates uptake of nutrients.

The choice of host is a compromise. The microorganism has to grow rapidly on a relatively cheap substrate, and ideally require no additional vitamins or growth factors. The host must synthesize the product in sufficient quantity and it must be easily harvested by downstream processing. In addition, for heterologous protein expression there has to be product integrity, e.g. protein folding and post-translational modification. Genetic stability is important, as is the ability for the host to be easily **genetically modified** with molecular techniques.

Industries using microorganisms will prefer to use those which are generally regarded as safe (GRAS). Examples of these microbes are provided in Table 1.1. Those which are on the GRAS list have been approved by the US FDA. Organisms which

Table 1.1 Examples of microbes generally regarded as safe (GRAS)

Microbe	Example
Bacteria	*Bacillus subtilis*
	Bacillus licheniformis
	Bifidobacter bifidum
	Bifidobacter infantis
	Lactobacillus bulgaricus
	Lactobacillus casei
	Lactococcus lactis
	Leuconostoc oenos
	Streptococcus thermophilus
Yeasts (single-cell fungi)	*Candida utilis*
	Kluyveromyces lactis
	Kluyveromyces marxianus
	Saccharomyces cerevisiae
Filamentous fungi	*Aspergillus niger*
	Aspergillus oryzae
	Penicillium roqueforti

Reproduced with permission from Michael J. Waites, Neil L. Morgan, John S. Rockey, Gary Higton, Industrial Microbiology: An Introduction, Microbiology & Virology, October 2001, Wiley-Blackwell.

are not on this list can be used, but greater control measures need to be put into place. Those organisms that have been genetically modified (GMO), require even more control processes to ensure that there is no unintended release of the GMO to the environment.

Obtaining or isolating an industrially useful microbe

There are major culture collections of microorganisms all over the world, which form a wonderful resource for choosing microbes for use in industrial processes.

- National Collection of Type Culture (NCTC)
- National Collection of Industrial Food and Marine Bacteria (NCIMB)
- National Collection of Yeast Cultures (NCYC)
- Collection of the International Mycology Institute
- American Type Culture Collection (ATCC)
- Japan Collection of Microorganisms (JCM)
- Deutsche Sammlung von Mikroorganismen und Zelkulturen
- Pasteur Culture Collection of Cyanobacteria (PCC)

However, these microbial collections may not contain the microbe which produces the metabolite which you want. Therefore it may be necessary to try and isolate your own microbial culture collection and screen it for the desired metabolite. The steps required to do this are outlined in Figure 1.3. Case study 1.1 provides an example of how important it was to isolate and develop new strains for the industrial production of penicillin.

Figure 1.3 Screening of a microbial culture collection.

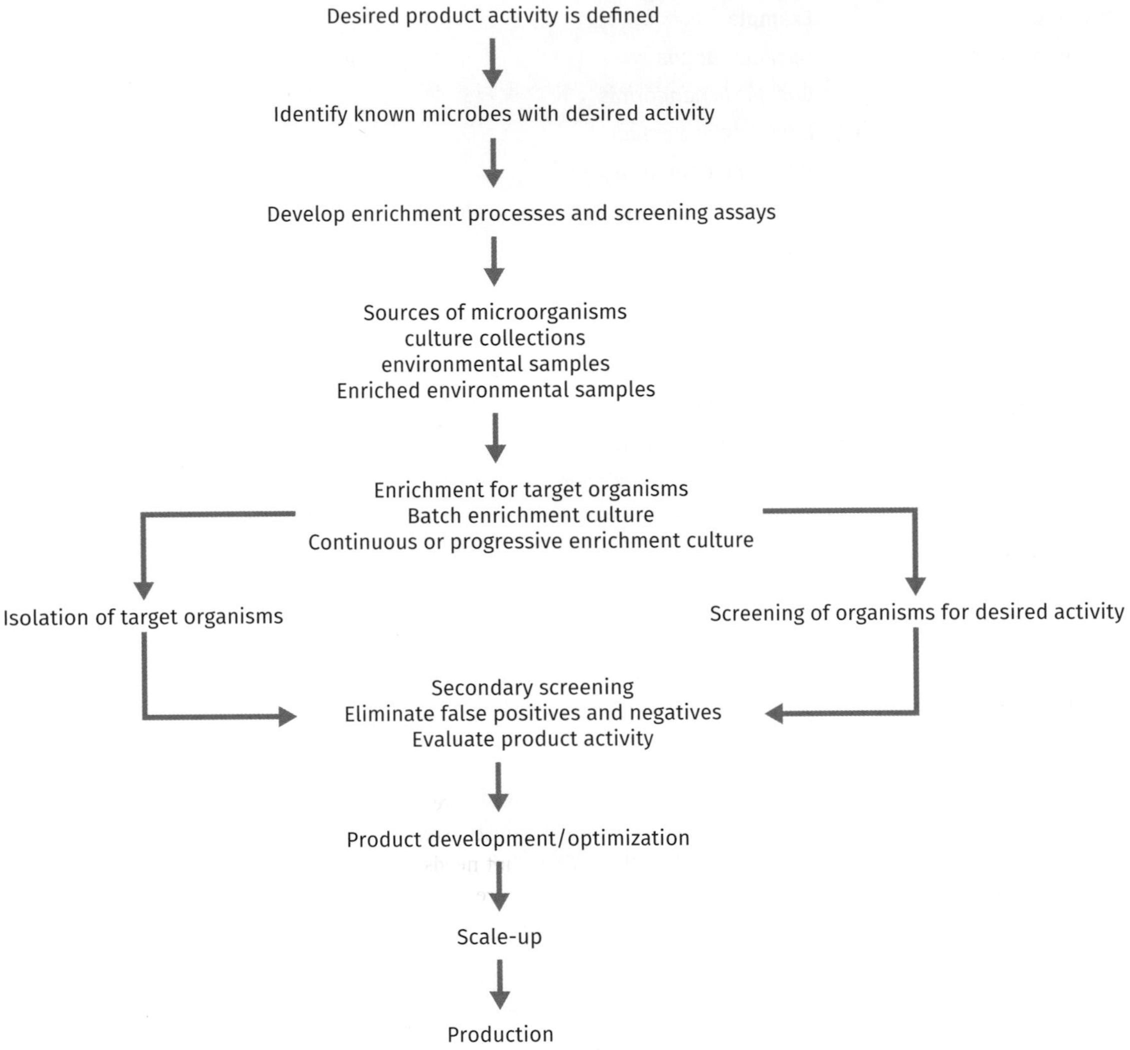

Strain development

Finding a microorganism which produces a promising **metabolite** (e.g. via a screening approach for antimicrobial compounds) can already be deemed a success. However, that microorganism will only produce specific quantities of this metabolite as is biologically necessary. Obviously, the point is to produce as much of the metabolite as possible to make the process a commercial success. The yield can be maximized by optimizing the cultivation conditions (e.g. nitrogen or oxygen limitation for antibiotic production) up to a certain point. Yet, biotechnologists aim for much higher productivity to maximize eventual profit from sales. This is where creating an 'over-producer' comes into play. An over-producer can be achieved by **random mutagenesis** or more specific site-directed mutagenesis.

Random mutagenesis inserts mutations randomly and is mainly used to change genomic information of the producer. Random mutagenesis uses a **mutagen**. Many mutants will have no apparently changed **phenotype** and this is due to the redundancy of the genetic code. Other mutant strains are not viable and will die. However, there should be a sufficient number of mutants generated with new phenotypes that can be screened for over-production or any other desired phenotype. Mutagenesis is often conducted within continuous culture to increase the number of mutants generated and thus increase the chances of finding an over-producer. The mutagen used in the process can be physical such as:

- UV irradiation (causing pyrimidine dimers);
- gamma irradiation (causing chromosome breakage);
- X-ray irradiation (causing breakage of DNA).

Mutagens can also be chemical such as:

- nitrous acid which causes deamination of bases;
- ethyl methane sulphonate or nitrosomethyl guanidine which causes alkylation of bases;
- base analogues (e.g. 5-bromouracil, 2-amino-purine) which are incorporated by mistake into DNA;
- intercalating agents (e.g. acridine orange) which intercalate within the pairs of bases in the DNA structure. This can then favour insertions or deletions of bases during DNA replication which can then lead to frameshift mutations.

Mutations can be randomly inserted into the chromosome by **transposons**, which are mobile genetic elements. In plasmids, mutations are introduced via PCR-based random mutagenesis by *Taq* polymerase which has a high error rate of up to 2 percent due to its lack of 3' → 5. exonuclease activity. PCR-based random mutagenesis is used particularly for vectors such as **plasmids**, because they are small enough to be produced as an amplicon and a library can be established quickly for screening.

Site-directed mutagenesis of the producer genome is a much more targeted method of obtaining an over producer. However, it can only be used if it is already known where the metabolic production bottleneck is that needs removing, or for the vector. DNA shuffling is a suitable method where homologous recombination of slightly variant genes or sequence segments takes place.

Case study 1.1
The production of penicillin

You are probably familiar with the accidental discovery of penicillin by Alexander Fleming at St Mary's Hospital in London. However, the journey of penicillin from a laboratory curiosity to a major pharmaceutical product required not only an understanding of how to purify it in sufficient quantities, but the development of a better strain for production and the ability to grow it on an industrial scale.

When Howard Florey, Ernst Chain, and Norman Heatley began to develop the purification process for penicillin they realized they needed to grow the *Penicillium notatum* fungus on a large scale. Unfortunately, *P. notatum* would not grow in submerged culture. Norman Heatley designed a culture vessel and the Staffordshire factory of James MacIntyre & Co were asked to produce ceramic culture vessels to enable greater production capacity. These culture vessels, with their inoculation and harvesting necks, can be seen in the Case study Figure CS 1.4. Despite being able to increase the culture volume, it was not enough to obtain a sufficient quantity of penicillin.

Large-scale production was not going to be possible in the UK due to Second World War bombing. Florey used his contacts in the US to secure production at the Northern Regional

Figure CS 1.4 Ceramic penicillin culture vessels.

The Sir William Dunn School of Pathology, University of Oxford. MS. 12202. Photogr. 1, Negative 33: Penicillin girls Bodleian

Research Laboratory in Peoria, Illinois. This research facility specialized in fermentation technology and had crucial knowledge about *Penicillin* spp. The location of the facility in the American Corn Belt was also important—it meant that corn steep liquor could be added to the fermentation, which increased the yield of penicillin. The next step was to get a better strain of *Penicillium* spp. The original species discovered by Fleming, identified as *P. notatum*, could only produce 4 units/ml of penicillin (1 unit = 0.6 μg). The search for a new strain was launched by importing soil samples and scouring the local markets for mouldy produce. Eventually, a cantaloupe melon arrived in the laboratory from which was isolated *Penicillium chrysogenum* which could grow in submerged culture and make 70–80 units per ml. A mutant was then isolated which could produce 250 units/ml. Screening ultraviolet light mutants of *P. chrysogenum* became a team effort involving many institutions across the US such as Stanford University and the Carnegie Institution. Using such mutants production was increased to 900 units/ml. Today's industrial strains, all derived from the *P. chrysogenum* isolated from the original mouldy cantaloupe, can produce 50,000 units/ml.

Key points

- Microbes have properties which make them very useful for biotechnology processes.
- It is important to use, where possible, microbes which are generally regarded as safe for production.
- Microbes can be obtained from many different environments, but culture collections also exist which are useful sources.
- Strain development allows increasing host productivity and/or activity of the product.
- Random mutagenesis or site-directed mutagenesis can be used for strain development.

1.3 Genetic manipulation

Genetic manipulation can further optimize an organism for use in a biotechnology process, for example by increasing production of a native compound or allowing utilization of a novel substrate. This section will provide an overview of genetic manipulation, while a more detailed explanation of the different methods and the issues involved with each is outlined in Chapter 6. If you examine Figure 1.5 you will observe some of the basic plasmid tools used in this process.

Plasmids, which are circular DNA strands, are integral to the majority of genetic manipulation methods. Typically, plasmids contain an *Escherichia coli* **origin of replication**, which allows for independent replication in this host, a **selectable marker**, usually an antibiotic resistance cassette, and a multiple cloning site, for insertion of fragments of DNA, typically genes required for protein expression. Some plasmids, termed 'shuttle' (Figure 1.5(a)), incorporate an additional origin of replication specific to the organism that the plasmid will be introduced into, allowing replication in this species. The origin of replication also determines the copy number of the plasmid, which can be as much as

Figure 1.5 Methods for genetic modification of microbes.

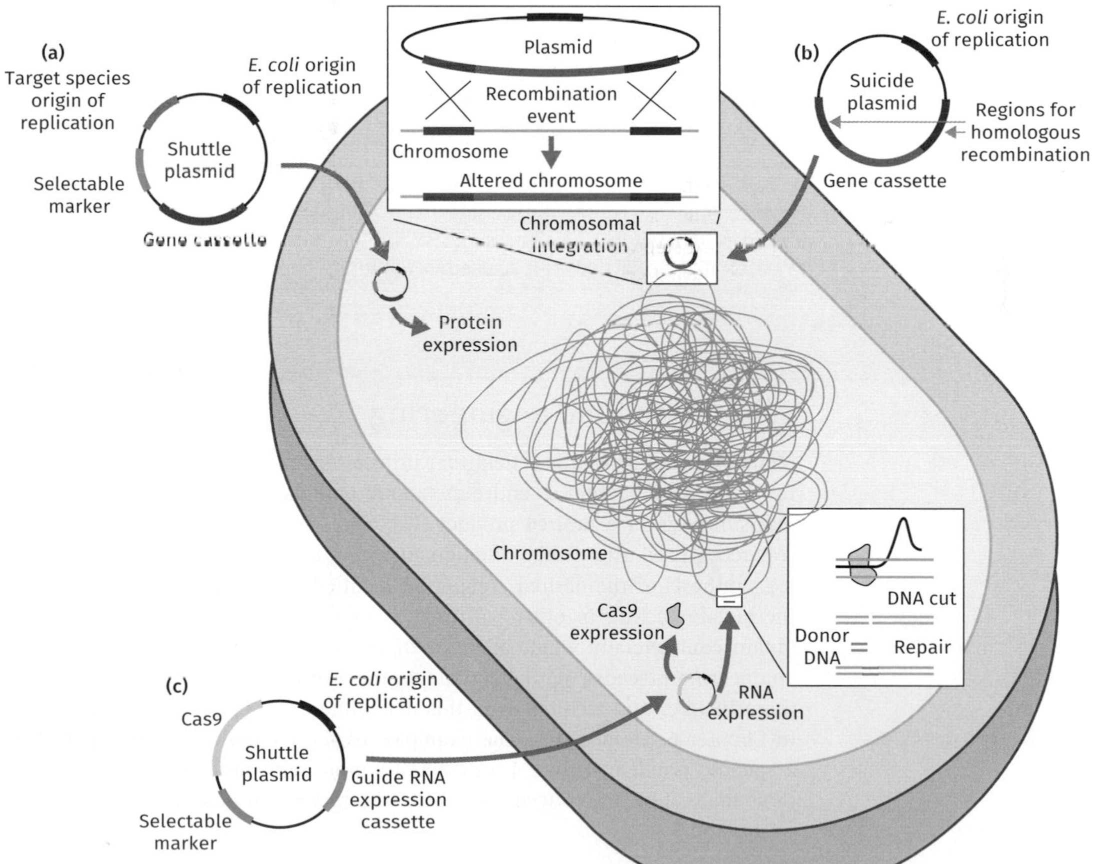

several hundred per cell, a significant advantage if high expression of a gene is required. Other plasmids, termed 'suicide' (Figure 1.5(b)), lack the target organism origin of replication and are used specifically to introduce foreign DNA into the chromosome by a process called **homologous recombination**. An example of a plasmid used in this process is one incorporating two fragments of DNA identical to specific regions in the chromosome, with an addition fragment of foreign DNA between them. Homologous recombination has the outcome of the identical regions on the plasmid and chromosome swapping over, or recombining. At the same time, the fragment of DNA between these regions is incorporated into the chromosome. The suicide plasmid, which is unable to replicate, is rapidly lost when the cell divides.

In many species, CRISPR (Clustered Regularly Interspaced Short Palindromic Repeats) based systems are becoming the tool of choice to modify chromosomes. It consists of two components. The first is a strand of RNA, termed a 'guide RNA', incorporating a specific sequence approximately 20 bp long, which in theory binds to a precise region of the chromosome. This guides the second component, typically an enzyme called Cas9, to this site, where it cuts the DNA. Both components are generally expressed using specifically designed plasmids (Figure 1.5(c)). The cellular machinery, recognizing the damage, automatically repairs the break. It is during this process that the chromosome can be modified by introducing DNA containing the specific modification required, which is incorporated into the chromosome by homologous recombination.

Key points

- Plasmids are key for genetic manipulation of microbes.
- Foreign DNA can be introduced into microbial chromosomes via homologous recombination.
- CRISPR or plasmids can be used to genetically modify organisms.

1.4 Metabolic engineering

Metabolic engineering is a field that utilizes genetic manipulation techniques for direct modification of entire metabolic pathways, in order to obtain optimal synthesis of a desired product. The necessity for metabolic engineering was recognized after initial attempts at introducing just the proteins required for synthesis of the desired compound resulted in low yields. Modulation of the metabolic flux in other pathways is generally also essential in order to obtain commercially viable amounts of the desired product. While metabolic engineering preceded synthetic biology, the two fields often overlap. Many of the principles and techniques in this discipline are discussed in greater detail in Chapter 6, along with some examples of the technology being applied to a specific problem. Figure 1.6 details a simple schematic showing modifications that can be introduced into a strain to improve production of the desired compound.

Figure 1.6 Modifications that can be introduced into an organism by metabolic engineering to increase product yields.

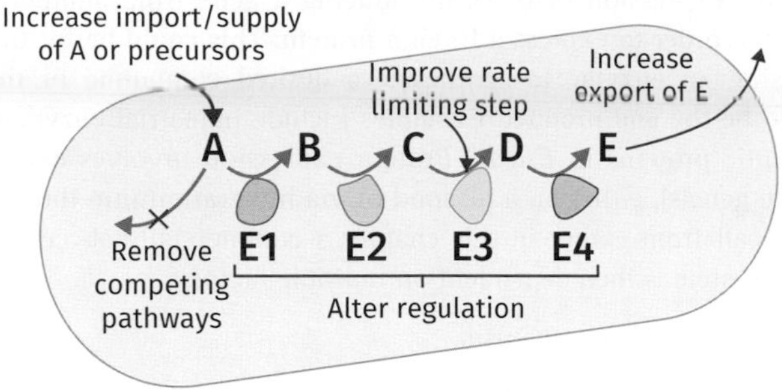

In Figure 1.6, a metabolic pathway consisting of four enzymes (labelled E1–E4), which could be either native or foreign, results in conversion of a substrate (A) to the desired product (E), via a series of intermediates (B, C, D). In theory, production of E could be increased by:

1. Introducing transport proteins either for A, precursors of A, or any of the intermediate metabolites.
2. Removal of other metabolic pathways that utilize A–D as a substrate could release more for production of E.
3. Modifying rate-limiting steps which occur when an enzyme(s) is slower in catalysing a certain reaction or found in lower quantities than the others in the pathway. Increasing enzyme amounts or modifying its catalytic properties can potentially resolve this issue.
4. Introducing proteins for export of the final product. This also prevents potential toxic accumulation within the cell.
5. Altering regulation of the pathway.

It should be noted, however, that the effect of these modifications may be limited if they divert energy and resources from growth to metabolite production. Cells that can reverse such changes will have a growth advantage, rapidly dominate a culture, and decrease compound production. In addition, these modifications require a host that can be repeatedly genetically manipulated and for which extensive knowledge of their metabolism is available—limiting the organisms that metabolic engineering can be applied to.

Key points

- Metabolic engineering involves genetically manipulating entire or multiple biochemical pathways.
- It is often required to increased product yields to commercial levels.
- Metabolic pathways can be modified in different ways to increase product yields.

1.5 Heterologous protein expression

Heterologous expression involves introducing a gene from another organism into a host in order to express a foreign protein. This could be for the purpose of expressing an enzyme to synthesize a desired compound or the protein itself could be the end product. Examples include industrial enzymes in fungi or therapeutic proteins in *E. coli*. Protein expression involves expressing the appropriate gene(s), either on a plasmid or via integration into the genome. As you will recall from earlier in this chapter a commercially successful protein expression system is then dependent on multiple factors:

1. Growth of the host organism.
2. Cost of the growth medium.
3. Whether the protein is retained in the cell or secreted into the growth medium.
4. Expression levels.
5. Toxicity of the protein to the host.
6. Protein folding.
7. Post-translation modifications like **glycosylation**.
8. Genetic amenability of the host.
9. Value of the final product.

For example, *E. coli* is easy to genetically manipulate and grows quickly but does not secrete or glycosylate proteins, making it unsuitable for expression of many eukaryotic proteins. Conversely, filamentous fungi such as *Aspergillus niger* are more difficult to genetically manipulate but secrete proteins, thereby reducing the cost of harvesting the desired product.

With certain proteins, cell toxicity is an issue, severely limiting production. To some degree, cell-free systems can overcome this issue since protein expression is uninhibited by lysed cells. As outlined in Figure 1.7, this system first involves obtaining a high-density culture of the desired organism, for example *E. coli*, which is then lysed, resulting in an extract containing the internal cellular contents and vesicles formed from membrane fragments.

Initiation of protein expression occurs by adding a plasmid or linear DNA containing a cassette expressing the gene(s) of interest to the lysate, or by introducing a plasmid into pre-lysed cells. In the introduced plasmid, the gene is under control of an **inducible promoter** that is switched on by a specific compound added to the cell lysate. The membrane vesicles contain intact

Figure 1.7 Schematic diagram detailing establishment of a cell-free-system.

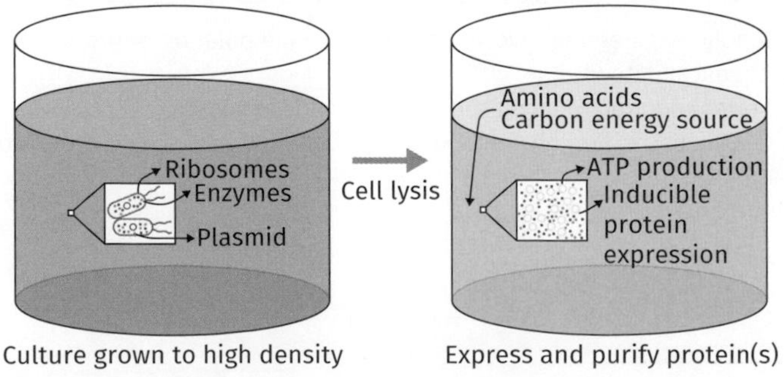

respiratory electron transport chains, allowing for production of ATP from an external carbon source. The various enzymes, metabolites, and ribosomes in the lysate are sufficient for protein production for a limited period, typically hours. The addition of amino acids increases production. Due to the limited production span, this method is generally only viable for high-value proteins.

Key points

- Heterologous protein expression requires a genetically modified host.
- Successful and commercially viable production is dependent on multiple factors.
- Cell-free systems can be used to produce proteins which are toxic to live cells.

1.6 Fermentation media and media design

The fermentation medium for the growth of a microorganism must contain all the nutrients it needs to grow, reproduce, and synthesize the maximum amount of the desired product. For example, aerobic microbial growth requires carbon and nitrogen sources, minerals, and oxygen. Water is also essential because all biochemical reactions are in aqueous solution.

There are many components which have to be added to fermentation media, not only carbon and nitrogen sources, but water, oxygen, vitamins, growth factors, and antifoams all have to be considered. All of these components are described in the following sections. It is important to recognize that media design is about balancing the cost of the substrates to make the media against the yield of the product and its market value.

Chemically defined fermentation media

A chemically defined medium is where all the components are known and defined, they are in a pure form, and exact amounts of chemicals are added to give known concentrations. Chemically defined media have consistency of performance and this type of medium is therefore important in the production of **biologics**, such as recombinant therapeutic proteins which are used in human treatments. The defined medium used to make human interferon-α_1 by *E.coli* is provided in Table 1.2 as one example.

Table 1.2 Medium for the production of human interferon-α_1 by *E. coli*

Component	Initial medium g/l	Feeding solution g/l
KH_2PO_4	3.0	
K_2HPO_4	5.0	
$(NH_4)_2SO_4$	4.0	30.0
$MgSO_4.7H_2O$	2.0	5.0
Vitamin B1	0.1	2.0
Trace metal	3.5 ml	
Glucose	30.0	500
Antifoam	0.5	
pH	7.0	

Other advantages of using this type of medium is that you can study how the product is made during the growth of the organism, and individual components can be varied one at a time to assess changes in growth and production. Downstream processing of the product might also be cheaper.

Complex fermentation media

In contrast to defined media, a complex medium has components within it which have not been totally characterized, and the concentration of certain compounds, e.g. sugars and amino acids, are not accurately known. Despite this, most fermentation processes use complex media. This is mainly due to cost, as the fermentation media can be as much as 50 percent of the operating cost of the process. Thus, renewable resources are desirable, particularly agricultural, dairy, and wood processing wastes. However, some substrates, such as lignocellulose waste, require more pre-treatment than others before they can be used.

The raw material must be available throughout the year. Seasonally available substrates should be avoided unless they can be stored in a suitable form in sufficient quantity so that a fermentation process can occur throughout the entire year. Other considerations include transport costs and suitability of the substrate for moving by bulk handling equipment. Disadvantages of using a waste product as a substrate is that its production will not normally be increased to meet additional demands—also as the usefulness of a waste product increases, it acquires a higher value which can be problematic from an economic perspective. Table 1.3 provides some more examples of constraints and challenges. These challenges arise both in the laboratory and when the process is scaled to an industrial level. However, it needs to be noted that those problems which arise on an industrial scale are usually more difficult to solve than those which occur in the laboratory.

Table 1.3 Constraints and challenges

Laboratory	Industrial scale
Development time, e.g. new media development and methods for product optimization	Availability of raw substrate material throughout the year
Cost of development efforts	Transport costs of media substrate components
Lack of shaker space for, e.g. culture flasks	Batch-to-batch variability of complex media components
Precipitation reactions between media components	Media costs and price fluctuations of media components
Water quality for media making	Stability of supply company for media components
Dispersion of solid media components	Bulk storage and handling of media components
Effect of media components on assay techniques	Pest problems, e.g. insect or microbial contamination of media components
Effect of components on downstream product purification	Effect of media components on broth viscosity or power consumption
Foaming during the fermentation	Disposal costs of spent media
	Dust hazards linked to media components

Carbon and nitrogen substrates for microbial fermentation

A carbon source is required for growth, reproduction, production formation, and cell maintenance. Carbon requirement can be determined from the biomass yield coefficient (Y), which is a measure of the efficiency of conversion of the substrate into microbial cell material.

Worked Example

Determination of the biomass yield coefficient (Y)

$$\text{Y carbon } (g/g) = \text{biomass produced}(g)/\text{carbon substrate utilized } (g)$$

The yeast *Saccharomyces cerevisiae* was grown in a chemically defined medium containing 750 g of glucose. A total of 400 g of *S. cerevisiae* biomass was produced. What is the biomass yield coefficient?

$$Y = 400g/750g$$
$$= 0.53$$

If the biomass yield coefficient of the yeast *Candida utilis* when grown on glucose is 0.32 and you used 100 g of glucose in a chemically defined medium, what amount of biomass (g) would you expect?

$$0.32 \times 100g$$
$$= 32 \text{ g of biomass}$$

Nitrogen is an important element needed to build nucleic acids and proteins. The nitrogen sources for microbial fermentation media come from diverse industries:

- from agricultural products;
- from the brewing industry;
- from fish and meat by-products.

Examples of carbon and nitrogen substrates and how they are used in fermentation are shown in Table 1.4.

Oxygen and water

Oxygen requirement will be vary depending on the microorganism being cultivated. It is most often supplied as sterile air, which will contain 21 percent (v/v) O_2, but some processes do require pure O_2. The air is sterilized through filtration devices. Water is the major component of media where submerged fermentation occurs, but it also has other uses in the industrial process for cleaning, heating, and cooling. Unlike in laboratory conditions, the water used in industrial fermentation media is not distilled or deionized. Fermentation processes will use water from rivers, lakes, or wells—so the water may contain dissolved metals and other impurities. It will also vary in pH and possibly have effluent contamination. Water quality is an important variable, particularly for some processes such as the production of beer. Water has to be recycled where possible as thousands of litres can be used per day. Hard water has to be treated to remove salts such as calcium carbonate. For some industrial processes, such as

Table 1.4 Carbon and nitrogen substrates used in industrial fermentation processes

Substrate	Description	Carbon/nitrogen	Example of use
Molasses	The sugary residue of primary sugar recovery	Carbon	Production of citric acid, single-cell protein (SCP), and antibiotics
Bagasse	The fibrous cane left over from sugar extraction	Carbon	Solid-state fermentation to produce penicillin
Fruit pomace	The solid remains after fruit juice has been extracted	Carbon	Ethanol production
Whey	By-product of the cheese industry, rich in lactose	Carbon	Production of SCP and propionic acid
Cassava, corn steep liquor, wheat processing waste	These are agricultural waste materials rich in starches	Carbon	Production of enzymes and alcohol
Sulphite waste liquor	Waste material from the paper-making industry after wood is digested to cellulose pulp	Carbon	SCP and ethanol production
Cellulose	The sources include trees, agricultural crop residues, and municipal waste	Carbon	Solid state fermentation for mushroom production
Hydrocarbons	Petroleum industry	Carbon	SCP
Fats and oils	Sources of fats and oils include fish, lard, soya bean, sunflower, peanut, corn, and cottonseed	Carbon	Antibiotic production
Oil seed meals	Proteinaceous materials from agricultural waste consist of oil seed meals, including soya, sunflower, and rape	Carbon and nitrogen	Enzyme and antibiotic production
Brewing industry waste	Yeast cells (used to ferment the wort in brewing) used as a supplement rather than main source of nitrogen	Nitrogen	SCP
Animal waste	Protein-rich meat and fish meal	Nitrogen	Antibiotic production

the production of biologics in the pharmaceutical industry, the water must be pure and is often filtered.

Growth factors and minerals

Growth factors are compounds not synthesized by the microorganism, but are required for growth, so they must be provided in the medium. Growth factors include vitamins and co-factors of enzymes. They can be very expensive to buy in a purified form and in sufficiently large quantities, so it is cheaper to ensure that the complex medium used in the process contains them. Some of the sources for vitamin growth factors can be found in Table 1.5. Alternatively,

Table 1.5 sources of growth factors

Growth factor	Sources
Vitamin B	Rice polishing, wheat germ, yeasts
Vitamin B2	Cereals, corn steep liquor
Vitamin B6	Corn steep liquor, yeasts
Vitamin B12	Liver, silage, meat
Nicotinamide	Liver, penicillin spent liquor
Panthothenic acid	Corn steep liquor

pathways for production of growth factors can be genetically engineered into organisms, if tools are available (see section 1.3).

Minerals such as iron are needed for growth, metabolism, and sometimes pH control, and can be required in trace amounts or as major nutrients. Many of the organic substrates used in fermentation media will contain some of the minerals needed. For example, corn steep liquor is known to contain a wide variety of minerals. Supplements are often provided such as potassium and magnesium-containing compounds. Often water impurities will also provide essential minerals.

Precursors, inducers, and inhibitors

Precursors are used to increase the yield or the quality of the product. They are needed in certain processes, such as in the production of antibiotics, and are generally provided through complex substrates such as corn steep liquor, or by the addition of pure compounds. For example, D-threonine is added to the fermentation media of the bacterium *Serratia marcescens* to produce L-tryptophan. Another example is the addition of phenylacetic acid, a side chain precursor, which is needed in penicillin production. If a product requires an inducer before it can be made by the microorganism, then this must be added to the medium. A good example is the production of enzymes, which will need the inducer present, i.e. the enzyme substrate before it will be produced. For example, amylase will only be produced when starch is present. Inducers are often needed for production in genetically modified microorganisms, as expression of the cloned gene can inhibit growth. The inducer is added to the fermentation media after growth has been established, switching on gene expression. A good example is commercial production of human interferon $\alpha 2$, using the filamentous fungus *Aspergillus nidulans* where growth of the fungus is allowed to occur until a sufficient biomass is reached, then ethylmethylketone is added which induces the production of interferon $\alpha 2$.

Inhibitors are also used, as they can push metabolism down a particular pathway and stop the formation of undesired intermediates. A good example is the production of glycerol by the yeast *Saccharomyces cerevisiae*, where the inhibitor sodium bisulphite is used to over-produce glycerol by trapping acetaldehyde. This works as the trapped acetaldehyde is then unable to serve as an electron acceptor for cytosolic NADH. The accumulated NADH is instead oxidized by the reduction of dihydroxyacetone phosphate to glycerol-3-phosphate which is the substrate for glycerol production.

Antifoams

These are media components, such as polypropylene glycol and silicone emulsion, which are needed to reduce the foam produced during a fermentation process. Foaming comes from the media proteins that become attached to the air–broth interface. Excess foaming can impair the monitoring process of the fermentation which can then affect the quality of the product. Antifoams are surface-acting agents that reduce surface tension and destabilize protein films. They can be added in one go at the start of the fermentation process, or may need to be added slowly throughout the process.

Media sterilization

It is essential that the media for some fermentation processes are kept sterile and that only the microorganism needed to make the product is being cultivated. This is important for processes such as antibiotic production, but not, for example, cheese manufacture. The fermentation medium is sterilized inside the bioreactor using a batch or continuous process. The batch process is done at 121°C for 20–60 minutes by injecting super heated steam directly into the medium, or by injecting steam into interior coils. The entire batch process can take several hours for heating and subsequent cooling leads to and other disadvantages such as media discolouration. The alternative to the batch process is continuous sterilization where the steam is injected into three different heat exchangers. The first heat exchanger rapidly raises the temperature to 90–120°C within 30 seconds, the second raises it to 140°C for 30–120 seconds, and the final heat exchanger cools it down. Other components coming into the bioreactor, such as oxygen, are filter sterilized.

Key points

- The fermentation medium is a key cost component of the industrial microbial process.
- Chemically defined as well as undefined complex media are used in industry, but the latter is more common.
- Agricultural and other industrial waste materials are good sources of both carbon and nitrogen.
- Other additions to the medium have to be considered, for example, the addition of vitamins, minerals, and precursors.
- An industrial process has to consider other issues such as water quality and the supply of oxygen, as well as the effect which foaming of the fermentation can have on the process.

1.7 Types of reactor

Reactors or fermenters are tanks and vessels in which cells are grown in the media. Examples of reactor types and the products made in them are provided in Table 1.6. All bioreactors have a working volume of the media, microbes, gas bubbles, and a **headspace**. The working volume is typically 70–80 percent of the total reactor volume, but more headspace might be needed if fermentation produces a lot of foam. The bioreactor operation mode is either a **batch process**,

Table 1.6 Bioreactor types and the products which can be produced in them

Reactor type	Product made
Stirred tank reactor (STR)	Antibiotics
	Citric acid
	Exopolysaccharides
	Various enzymes, e.g. cellulose and xylanase
Bubble	Algal culture
	Chitinolytic enzymes
Air lift	Antibiotics
	Chitinolytic enzymes
	Single cell protein (e.g. Quorn)
	Exopolysaccharides
	Various enzymes, e.g. cellulose and xylanase
Fluid bed	Laccase
Packed bed	Laccase
Membrane bioreactor	Antibiotics
	Hydrogen production
	Water treatment

fed-batch process, or **continuous process**, and these operation modes can be used in either **submerged**, liquid culture, or **solid state** fermentation. Bioreactor operation modes are covered in more detail in Chapter 2 when we examine microbial growth kinetics. All bioreactors have control and measurement systems. These systems will measure the speed of the **impellers**, the temperature of the fermentation, the control of gas supply, the measurement and control of pH, the measurement of dissolved oxygen, antifoam control, and finally feed control.

Submerged reactors

Submerged reactors, which have no mixing or aeration, are used for the manufacture of traditional products such as beer, wine, and cheese. Most (70 percent) of the bioreactors used for microbial cultivation on an industrial scale fall into this category.

Stirred tank reactors and continuous stirred tank reactors

Stirred tank reactors (STR) and continuous stirred tank reactors (CSTR) are the most widely used systems for submerged culture of microbes which produce more modern products, such as antibiotics and enzymes. They are also a good scale-up design from the laboratory to production. STRs are commonly used for batch and fed-batch culture, whereas CSTRs are used in continuous culture, such as a **chemostat** or a turbidostat. Both types of reactor come in a variety of sizes from as small as 10 ml to 2,000 litres. An example of a stirred tank reactor is shown in Figure 1.8. They are cylindrical vessels, made of glass or stainless steel, with a rotating impeller system for mixing the culture located either at the top or the bottom of the vessel. The motor-driven central shaft can have more than one impeller, depending on the volume of liquid in the vessel. Swirling of liquid

Figure 1.8 Stirred tank reactor.

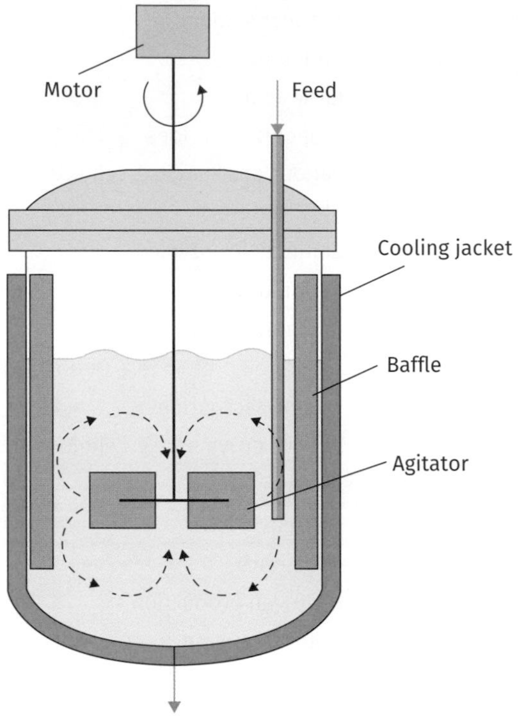

is a real issue, think about when you are stirring sugar into coffee—when you stir quickly the liquid rises up the inside of the cup. In bioreactor vessels **baffles** are often used to combat this issue. These are vertical strips of metal mounted against the wall of the tank which help to reduce vortexing and swirling and aids aeration. Gas comes into the vessel from below the mixing impeller. The main limitation of both the STR and the CSTR is the shearing stress produced during stirring. Stirring is essential to help the transfer of nutrients, heat, and oxygen through the fermentation. It removes gradients and also stops the build up of toxic products. However, stirring produces shearing forces due to difference in the velocity gradients of the liquid in the vessel. These forces are damaging to the microorganisms and also to the products. Shearing forces are particularly bad for filamentous fungi as the **hyphae** can become tangled and damaged.

Bubble column reactors and airlift bioreactors

Bubble column reactors (BCR) and airlift bioreactors (ARL) are examples of pneumatic (air) mixed reactors used for the culture of filamentous fungi, animal, and plant cells. A BCR is a cylindrical vessel with a gas distributor at the bottom. The gas is sparged in to produce bubbles in the liquid. BCRs are used for producing enzymes, proteins, and antibiotics—they are good at growing filamentous fungi. The airlift bioreactor (ARL), such as the example shown in Figure 1.9 is a variation of the BCR. It is used to culture organisms which produce antibiotics, as well as the production of microbial biomass for single-cell protein production (see Table 1.6). In an ARL a central tube is responsible for

Figure 1.9 Airlift fermenter.

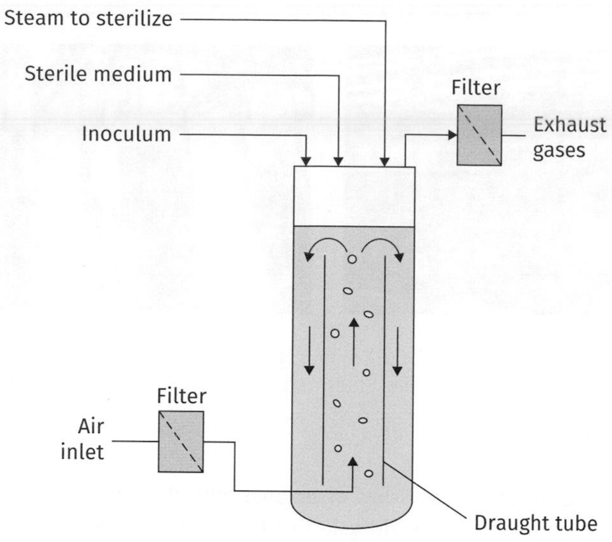

mixing and circulation of the fluid. The liquid circulation occurs as only the part of the liquid in the reactor located in the central draught tube is sparged with the gas to produce bubbles. The bubble-filled liquid has a lower density than the bubble-free liquid and the difference in the hydrostatic pressure forces the bubble-filled liquid to move upwards through the draught tube to the headspace in the reactor. In the headspace the excess air and CO_2 are released and the degassed liquid flows back down the outside of the draught tube. In BCRs and ARLs there is no mechanical stirring, thus they are vessels that have low shearing forces. However, there are limitations—there is back mixing between the gas and the liquid phases, high pressure drops, and also bubble coalescence.

Fluidized bed and packed bed bioreactors

Fluidized bed reactors (FBR) are tall column reactors. They have applications in wastewater treatment but are also used for microbial metabolite and biomass production. The fluidizing medium can be a gas, liquid, or a mixture of both.

A packed bed reactor (PBR) is, as its name suggests, comprised of a packed bed that supports cells either on or within carriers, and a reservoir used to re-circulate nutrients and oxygen through the bed. In a trickle-bed reactor, liquid is sprayed on top of the packing which trickles down the bed in rivulets.

Photobioreactors

Photobioreactors (PBR) are used for the cultivation of phototrophic microbes such as microalgae and *Cyanobacteria*. Microalgae are important in biotechnology and have a wide range of commercial applications ranging from pharmaceuticals, fish food, neutraceuticals, and as a fertilizer for the agricultural industry. Methods of cultivation vary and include open ponds and flat plate bioreactors, as well as tubular and vertical columns. Open ponds (see Figure 1.10(a)) require relatively low technology, but do need large areas of land. This method was developed from ideas around artificial lagoons and the oxidation ponds used in wastewater treatment. The shape of the pond varies from large, shallow, circular ponds to the raceway pond design shown in Figure 1.10(a). The raceway pond is rectangular, and paddles circulate liquid to ensure a

Figure 1.10 (a) Open pond bioreactor; (b) closed PBR horizontal tubes; (c) a vertical system for the growth of phototrophic microbes.

(a) (b) (c)

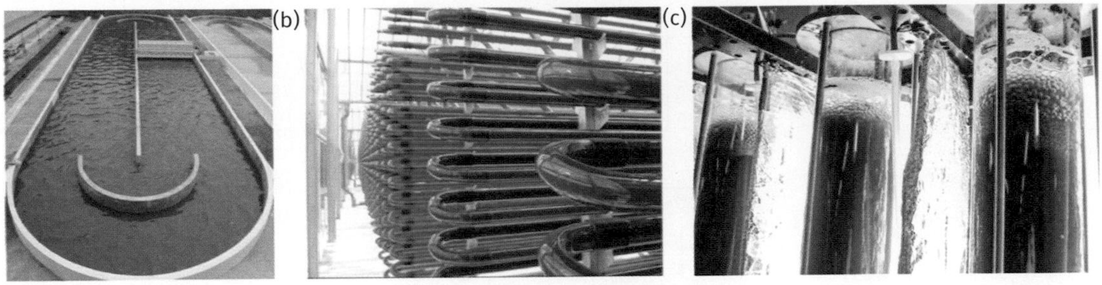

(a) Reproduced with permission from Satpati, Gour & Pal, Ruma. (2018). Microalgae- Biomass to Biodiesel: A Review. 9. 11-37. J. Algal Biomass Utln. 2018, eISSN: 2229 – 6905 (b) AJCespedes/Shutterstock.com (c) Eva Decker / Wikimedia Commons / CC-BY-SA-3.0

continuous flow. There are issues associated with this method of cultivation, including evaporation losses, diffusion of CO_2 to the atmosphere, exposure to variable weather conditions, contamination, and grazing.

Closed PBR systems are designed to overcome some of the issues which arise with open ponds. In a flat plate reactor the microbes are cultivated on large transparent plates which generally cover one side of a panel, which is tilted towards sunlight. A pump circulates the algal cells. Horizontal tubular photobioreactors are the most popular closed system, and one of these is shown in Figure 1.10(b). The tubes are arranged in different orientations. The tubes have diameters of 10–60 mm and can be up to several hundred metres in length. This is necessary to achieve a high surface area to volume ratio. Vertical systems operate in a similar way to bubble reactors and airlift fermenters except that the vessels themselves have to be transparent to allow light to pass through (Figure 1.10(c)).

Solid state fermentation

The alternative to submerged culture is to use solid state fermentation (SSF), where fermentation occurs without additional water. This system is particularly good for the growth of filamentous fungi and the largest industrial process to use this method is the cultivation of mushrooms. There are several different types of SSF bioreactors:

1. **Static bioreactor** consisting of a fixed bed or perforated trays. Fixed-bed systems are closed and aeration is forced. They can be in the form of columns or perforated trays. Perforated trays are a very straightforward system where the substrate for growth is packed into trays, which can be wood or stainless steel. The perforations allows air movement. The trays are then kept in a room where the temperature and humidity can be controlled. Mushrooms are produced on an industrial scale in this way (see Figure 1.11).
2. **Stirred bioreactor** consisting of a horizontal drum or where the drum is stirred. Rotating drums have been used for SSF since the 1930s. The drum agitation can be either continuous or intermittent and will often have baffles to allow for efficient mixing. This system is often used for bioremediation.

Figure 1.11 Commercial mushroom production.

VicPhotoria/Shutterstock.com

Key points

- There are many different types of vessel which can be used as bioreactors to grow microbes.
- Bioreactors operate as batch, fed-batch, or continuous processes which can be used in submerged or solid-state fermentation.
- Submerged reactors used for making traditional products such as cheese and beer are the most common.
- Stirred tank reactors are more commonly used to make pharmaceutical products.
- Airlift fermenters are good for growing filamentous fungi.
- Photobioreactors are used to grow *Cyanobacteria* and microalgae.
- Solid-state fermentation systems include static trays or rotating drums.

1.8 Scale-up and optimization

The development of a biotechnology process at the laboratory scale serves as a proof of concept. The process then has to be scaled-up (see Figure 1.12), and the large-scale process may need to be validated.

Scaling-up is not just a trivial increase in size and volume, but poses challenges which can result in loss or major reduction in product yields. Some of these challenges are merely technical (such as the need for several rounds of pre-culturing to get enough material to inoculate an industrial scale fermenter), whereas others affect the biological functions of the host system.

If we have high cell densities, the culture behaves very differently. Figure 1.13 shows how impressively the biomass can be increased in a fermenter in comparison to batch culture.

Although individual microorganisms are invisible to the naked eye and their mass is negligible, the biomass of all microorganisms produced in large-scale culture is that of human beings many times over. The production of heat by

Figure 1.12 Scale-up context within bioprocess development.

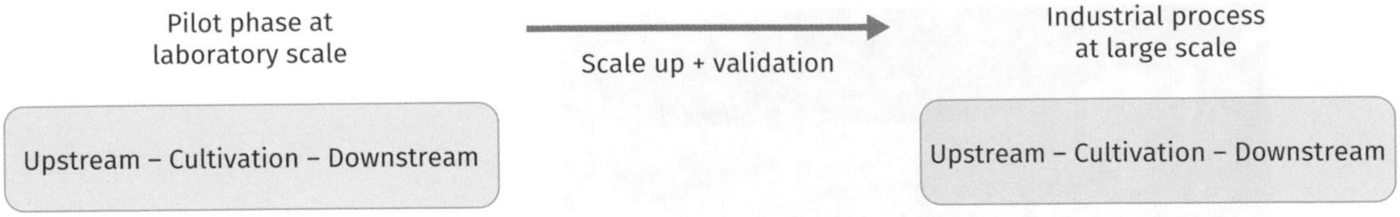

Figure 1.13 High cell density culture of *P. pastoris.* Left centrifuge bottle shows the cell sediment of a batch culture at a density of 1 at OD_{600}. On the right is the sediment of a culture grown in a fermenter at a density of 130 g dry cell weight per litre (about 500 at OD_{600}).

Joan Lin Cereghino, James M. Cregg, Heterologous protein expression in the methylotrophic yeast *Pichia pastoris*, FEMS Microbiology Reviews 2000; 24 (1): 45–66, doi:10.1111/j.1574-6976.2000.tb00532.x. Reprinted and translated by permission of Oxford University Press on behalf of Federation of European Microbiological Societies

a large-scale culture poses a challenge, and fermentation vessels have to be cooled. As cells grow within the fermentation media there is an increase in viscosity of the liquid. The high viscosity results in technical challenges for oxygen transfer and nutrient supply, as well as disposal of metabolic waste products such as acetic acid lowering the pH. Increased stirring results in shearing and production of foam, which impacts productivity.

Let us consider how large-scale affects plasmid stability. Plasmid stability is essential if the recombinant protein is encoded on plasmids with in the host cell. The production of the protein product has to be optimized and the plasmid maintained inside the host cell. Stability can be addressed in upscale development by optimizing plasmid structure, copy number, expression level, and the introduction of stabilizing sequences. The physiological factors impacting on plasmid stability depend on growth phase and rate, which in turn depend on the cultivation medium, oxygen availability, temperature, and pH.

If this all goes well the biotechnologist is faced with the challenges of removing thousands of litres of culture medium, e.g. by filtration or centrifugation. Sometimes gravity-based cell settlers are used, and sometimes centrifuges, or alternating tangential-flow filters. These procedures are effective at laboratory and large scale. However, separating such large volumes takes time, requiring everything to be cooled to avoid/limit degradation and lysis of cells and the product, or loss of product activity.

Deciding when to harvest the culture and how swiftly all downstream process steps take place also affects how much host cell protein contaminates the product—thus its eventual purity.

Key points

- Scaling up from laboratory to industrial process is a challenge and often results in loss of the product.
- The challenges of high cell density cultures are related to heat production and high viscosity.
- Physiological factors affected by culture conditions impact on plasmid stability.
- Large-scale harvest requires specialist equipment and stresses the cells, which affects product quality.

1.9 Downstream processing

We have now logically and procedurally arrived at the downstream development. As laid out in Figure 1.14 the steps taken depend on what kind of product we aim for and if the product remains in the host cell or is secreted into the medium, whether it is soluble or not, and if it is active or not.

Figure 1.14 Downstream development steps.

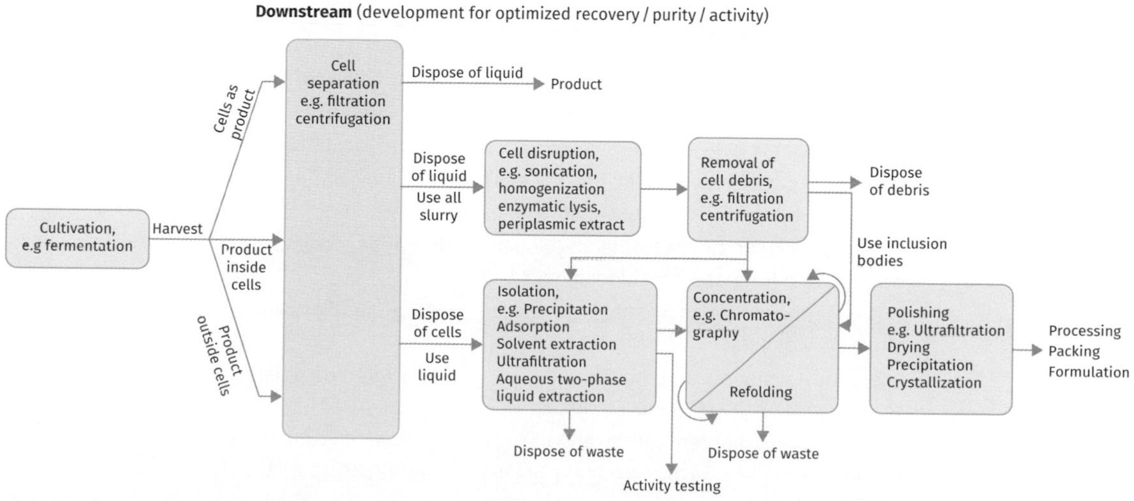

Cell separation is often achieved by filtration or centrifugation. If cell disruption is necessary, sonication with high-frequency waves or physical homogenization (e.g. using pressure or friction), as well as enzymatic lysis (e.g. using lysozyme) is an option. If the product is accumulated in a certain cell compartment, this compartment can be specifically targeted such as when performing a **periplasmic extraction**. Isolation of the product can be based on precipitation, adsorption, solvent extraction, **ultrafiltration** (membranes separate 1–100 nm molecules), or aqueous two-phase liquid extraction (two different immiscible water-soluble phases allow separation depending on, e.g., ionic strength, pH). Chromatography is usually applied for concentration steps. This can be **anion, cation exchange** and/or **interaction hydrophobic chromatography** performed as a flow-through process almost resembling ultrafiltration, or as a bind-and-elute process. Protein refolding may be necessary, e.g. if the production regime focuses on **inclusion bodies** as an output. These need to be solubilized and then refolded correctly, for instance by using folding promoting agents.

The eventual polishing step can involve ultrafiltration, drying/lyophilization, precipitation (e.g. ammonium sulphate precipitation for proteins), and crystallization (e.g. for amino acids or industrial enzymes). The process aims for the product formulation with the highest stability of the active product and the longest shelf-life.

Key points

- Downstream processing steps depend on product activity and solubility and the type of host.
- Cell disruption may be necessary using sonication, homogenization, enzymatic lysis, or periplasmic extraction.
- The product can be purified and polished using precipitation, adsorption, solvent extraction, ultrafiltration, or chromatography.
- Protein refolding may be necessary.

Chapter Summary

- Any biotechnology process requires an upstream and a downstream phase.
- The development of a biotechnology process requires input from a range of different scientists.
- The host organism used for the process has to be chosen carefully. It is preferable for the organism to be GRAS.
- Strains of microbe can be further developed using genetic modification which includes the tailoring of metabolic pathways.
- Fermentation media for microbial growth can either be complex or chemically defined.
- There are constraints and challenges in media design when considering scaling up from the laboratory to industry.
- There are many different carbon and nitrogen sources which can be used in complex media design. Many of these are waste products from other industries.

- Consideration has to be given to other additions to the media, such as the water, oxygen, precursors, and inhibitors.
- Reactors for the cultivation of microbes can be operated using batch, fed-batch, or continuous culture, and these can be either submerged or solid state.
- Stirred tank reactors are most commonly used for submerged culture and photobiorectors are used for the cultivation of phototrophic microbes.
- Scale-up of a biotechnology process to industrial level can be challenging from both a technical and biological perspective.
- Downstream processing is essential in order to recover the desired product.

Further Reading

Demain, A. L. (2007). 'The Business of Biotechnology'. Industrial Biotechnology 3(3): 269.
For a review of the products which make up the biotechnology industry, this is a useful and readable review article.

Singh, V., et al. (2016). 'Strategies for fermentation medium optimization: An in-depth review'. Frontiers in Microbiology 7: 2087.
For an in-depth review on how to approach media design, this is an excellent source.

Volmer, J., Schmid, A., and Bühler, B. (2015). 'Guiding bioprocess design by microbial ecology'. Current Opinion in Microbiology 25: 25–32; https://doi.org/10.1016/j.mib.2015.02.002
This review gives some insights into bioprocess design.

Discussion Questions

1.1 Outline the stages that need to be considered when developing a biotechnology process.

1.2 If you were considering small-scale production of oyster mushrooms (a premium product), what type of fermentation system would you use and why?

1.3 Explain the kind of choices we need to make when selecting a recombinant host.

1.4 Discuss the challenges when scaling up from the laboratory to industry.

2 MICROBIAL GROWTH

Learning Objectives

- To appreciate that microbial cells can be measured in a number of different ways, but that all methods have their advantages and disadvantages;
- to understand that bioreactors used for microbial growth can be operated using batch, fed-batch or continuous culture;
- to learn that there are specific growth phases associated with microbial growth in batch culture;
- to learn that different metabolites can be produced at different specific growth phases;
- to learn that each bacterial cell produces two daughter cells through binary fission causing exponential growth;
- to learn how to model bacterial growth in bioreactors;
- to understand how yields can be calculated during bacterial growth;
- to appreciate the importance of productivity and maintainance in fermentation.

The usefulness of microbes in biotechnology applications stems from their ability to convert low-cost substrates into higher-value products. Examples of commonly produced and microbially derived products include foods such as bread, wine, and cheese; food additives such as xanthan gum and monosodium glutamate; medicines such as insulin and many antibiotics. You will find out more about these products and how they are made in Chapters 4 and 5.

You will remember from Chapter 1 that microbes are often grown in a liquid medium, which allows for control of growth through aeration, mixing, and supplementation. Microbes can also be grown using **solid-state** fermentation where the microbes grow on a solid nutrient. It is harder to control microbial growth in solid-state fermentation due to difficulties of mixing and aeration, but many important biotechnology processes use solid waste as a feedstock, for instance the fermentation of soy sauce from soy beans, the generation of biofuels such as biodiesel from food waste, and the growth of mushrooms in the food industry.

Key to optimizing production is understanding how microbes interact with and grow in the substrates provided, and how microbial metabolism is linked to product formation. In this chapter we will look at how microorganisms such as bacteria grow in liquid culture, the kinetics used to model bacterial growth and how yield of different products can be optimized through process control, such as dilution and concentration of nutrient. First we will examine how microbial growth is measured when grown in liquid culture.

2.1 Measurement of microbial growth

As we know from Chapter 1, microbial cells can be grown in liquid and solid media. It is important that we are able to monitor the growth of cells when grown in culture.

The number of cells in a population can be measured in a number of different ways and each method has distinct advantages and disadvantages. Some methods are more suitable for some organisms than others. In this section we are going to examine the use of total cell counts, viable counts, turbidity, and dry weight.

Total cell counts

A specialized microscope slide, e.g. a Neubauer or haemocytometer counting chamber which has a grid etched on the surface, is used for total cell counts. The slide is made of glass and the etched grid can only be seen under the microscope (Figure 2.1). A known volume of liquid can be trapped over the grid enabling cells to be counted per unit volume. Yeast cells and algal cells can be counted in this way, but bacteria, which are typically 1 – 2μm in size, are generally too small. Counting is quick, and this method can distinguish between dead and live cells if a stain such as trypan blue is used. Read Applying the concepts 2.1 to get an understanding as to how a haemocytometer is used.

Figure 2.1 (a) Haemocytometer slide; (b) grid on the slide seen under magnification.

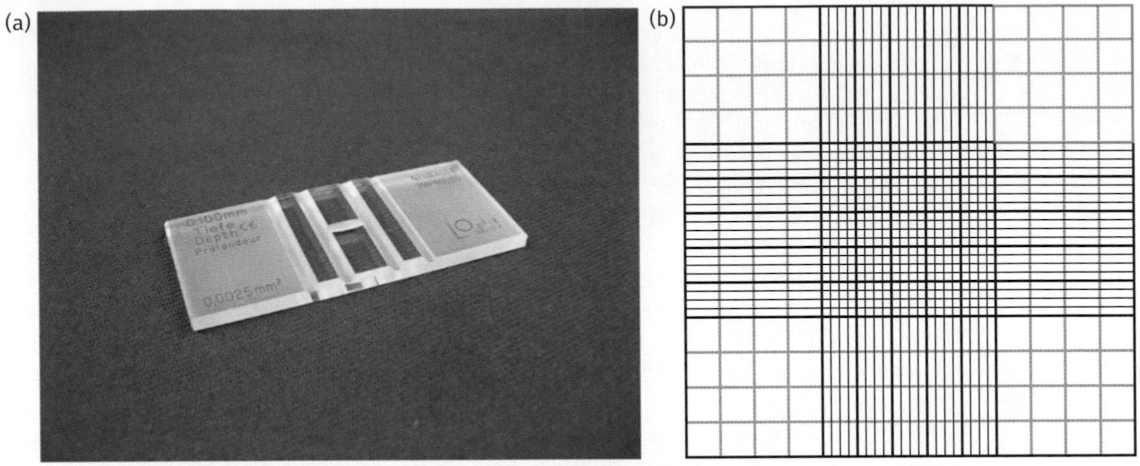

Applying the concepts 2.1
Counting cells using a Neubauer chamber

The grid of a haemocytometer contains nine squares (see the blue highlighted square in Figure 2.1(b)).The middle square is subdivided into 25 squares and each of these is then divided into 16 smaller squares. The suspension of the cells to be counted needs to be dilute enough so that the cells across the grid have an even distribution. The number of cells inside the four large corner squares and the middle square are counted. There is a method to the cell count: cells are counted which are found within the square or on the right hand, or bottom boundary lines. This method will not distinguish between live and dead cells unless a specific stain is used. For example, only dead cells take up trypan blue stain, so if you want to determine the number of viable cells, blue-stained cells would be discounted, or you could equally wish to enumerate the number of dead cells. When the cells have been counted, calculate the mean. To get a cell count per ml, the mean is then multiplied by 10^4. This is because the volume of liquid under each of the nine 'middle-sized' squares of the haemocytometer is 100 nl (or 0.1 µl, or 0.0001 ml). If the cells have been diluted with a stain prior to counting, then this dilution factor has to be taken into account when calculating the original cell count per ml.

Example

Cells in a suspension were diluted 1:5 with trypan blue. A sample was placed in a haemocytometer and the four outside squares and the middle square were counted. The counts were 50, 48, 54, 45, and 51.

1. Calculate the mean

   $(50 + 48 + 54 + 45 + 51) \div 5 = 49.6$

2. Multiply by 10^4

   $49.6 \times 10{,}000 = 4.96 \times 10^5$

3. Multiply by the dilution factor as the cells were diluted 1:5 with trypan blue.

   $4.96 \times 10^5 \times 5 = 2.48 \times 10^6$ cells per ml in the original suspension.

Viable counts

This method will determine the number of living cells in a solid such as soil, or a liquid such as sea water. It can also be used to monitor the growth of cells in broth culture. It relies on using a dilution series and plate culture. Each individual microbial cell when on the plate divides to form a visible colony which can be counted; this is shown in Figure 2.2. The cell number is then expressed as a colony forming unit (cfu). There are problems with this method—cells can clump together leading to an underestimation of population numbers. Only cells that actively grow on plates can be counted, and the method is quite time consuming. A viable count is better for determining numbers of bacteria or yeasts rather than filamentous fungi.

Figure 2.2 The establishment of a dilution series followed by plate culture to determine colony counts (a viable count).

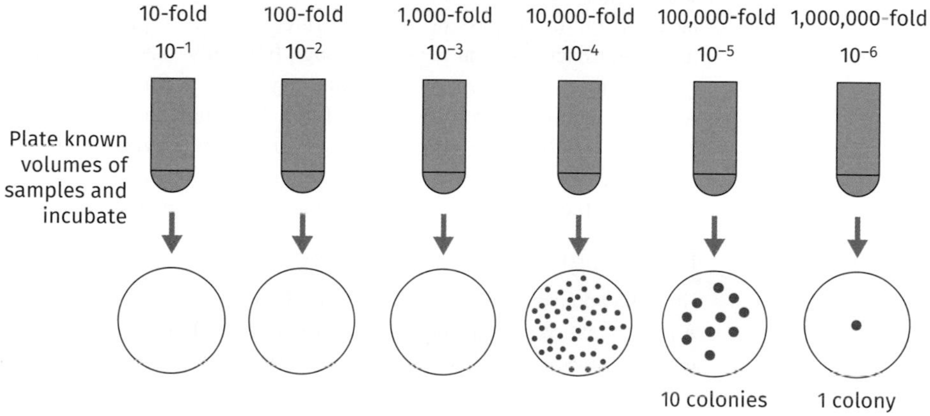

Applying the concepts 2.2

In an experiment to enumerate the number of *Streptomyces* bacteria in soil, you added 1 g of soil to 10 ml of sterile distilled water. A serial dilution has been performed from 10^{-1} to 10^{-6} and 100 µl of each dilution has been plated onto selective medium for the growth of *Streptomyces* spp bacteria. After incubation at 30°C, the number of colonies on each plate was recorded. You found that you obtained 60 colonies on the 10^{-3} dilution.

1. What is the colony forming unit (cfu) per ml of the *Streptomyces* bacteria in the sample?
2. How many *Streptomyces* are there in 1 g of the soil sample?

The 10^{-3} dilution is a 1000-fold dilution of the original sample. Therefore

$$60 \times 1000 = 6.0 \times 10^4$$

We know that 100 µl was spread onto the plate, so to get the number of bacteria in 1 ml (1000 µl) is

$$6.0 \times 10^4 \times 10 = 6.0 \times 10^5 \text{ cfu/ml}$$

The second question asks how many *Streptomyces* bacteria there were in the 1 g of soil. You know that 1 g of soil was placed into 10 ml of water. Thus, if the cfu/ml is 6.0×10^5, then

$$6.010^5 \times 10 = 6.0 \times 10^6 \text{ cfu/g}$$

Turbidity

The change in optical density of a clear liquid medium can also be used as a measurement of growth. This uses a spectrophotometer usually set at 600 nm. A beam of light is shone through the liquid and the reading depends on how much the light has scattered due to the number of cells in the suspension. The more cells, the more the light scatters and the higher the optical density (OD) reading. OD reading can be linked to viable counts or dry weight using a calibration curve. This method is quick, but it won't distinguish between live and dead cells. There is also the problem of particulate matter in the fermentation medium which may come from the substrate being used for growth that could increase the OD 600 nm reading.

Dry weight

Dry weight determination is a more lengthy procedure than turbidity measurements. A culture is passed through a filter paper with pore sizes small enough to prevent bacteria from passing through (typically 0.2 μm). The filter paper is typically dried in an oven at approximately 70–90°C for 24 hours and weighed. A known volume of culture is then filtered through the paper and re-dried at 70°C and weighed. The difference in weights then provides a dry weight measurement. This method will neither distinguish between live and dead cells nor is it useful if the media used for the cultivation has suspended particulate matter.

Measurements of macromolecules

Growth can also be determined using protein assays and assays for nucleic acid. A protein assay, such as the Bradford assay can only be used if the fermentation medium does not itself contain protein. DNA measurements can be used, but these require separation of the DNA and so are technically more complex than protein assays. A measurement of absorbance at 260 nm can be made and the concentration of DNA then determined using the molar extinction coefficient of DNA. DNA measurements are one of the few ways in which growth can be measured in media which has a high solid content.

Key points

- Growth of microbes can be measured in a number of different ways such as cell count and dry weight, but they each have their limitations.
- Measurements of cells using optical methods can be hampered by the particulate matter often suspended in the fermentation media

2.2 Microbial growth in different culture systems

When microbial cells are grown in both the laboratory and on an industrial scale they are cultured in containers known as **bioreactors**. These bioreactors are broadly split into the three different culture systems described in Chapter 1:

1. Batch culture
2. Fed-batch culture
3. Continuous culture.

There are distinct differences; look at the very simplified bioreactor diagrams in Figure 2.3. The **batch culture** is a closed system, medium is not added, or removed, thus the volume inside the reactor stays the same. Nutrients will be used up as growth occurs and waste products will accumulate. In **fed-batch culture** new nutrients in the form of fresh medium can be added at specific time intervals, but waste products will still accumulate as culture and spent medium is not removed. This is still a closed system, but the volume in the reactor does increase. In **continuous culture**, new nutrients are added in the form of fresh medium at a specific rate, and culture and spent medium is removed at the same rate, thus waste products cannot accumulate and growth is not impeded, the volume in the reactor does not change, and it is an open system.

Understanding and modelling the growth of microorganisms in different culture systems is an important tool in biotechnology. These culture systems provide different environments for growing microorganisms and the microorganisms respond to these environments in different ways.

We will consider how each of these culture systems can be modelled in turn. First, we will consider how microbes are grown in a batch culture bioreactor. Batch bioreactors are used in many biotechnology processes, for example the fermentation of beer and wine, and are the oldest and simplest culture systems in microbiology. Because of its simplicity, batch culture remains one of the commonest methods for the fundamental study of microbial growth.

Figure 2.3 Bioreactor diagrams showing three different culture systems.
(a) batch culture; (b) fed-batch culture; (c) continuous culture; (d) a photograph of a commercial bioreactor.

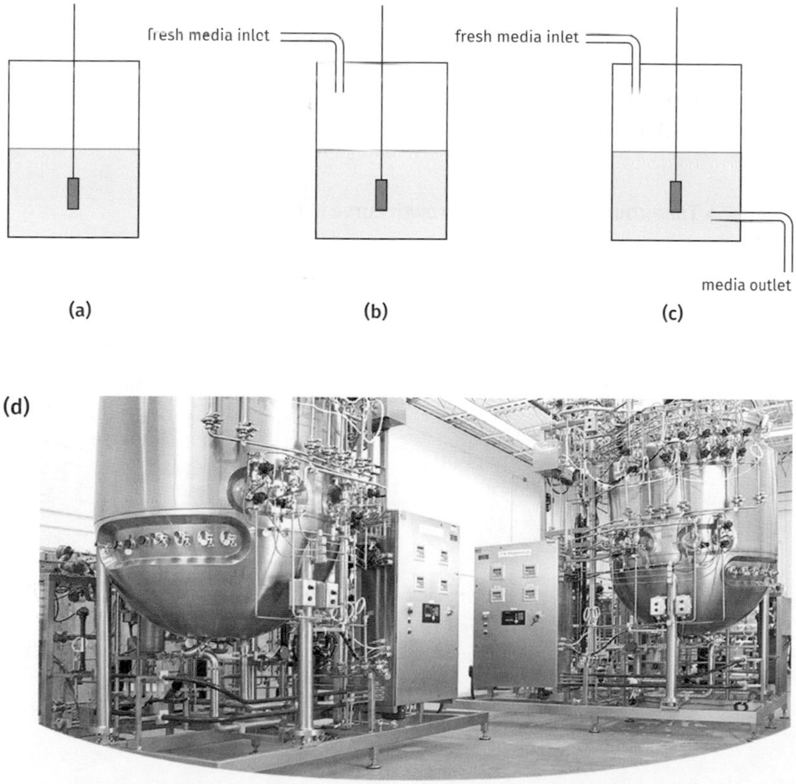

2.3 Microbial growth phases in batch culture

There are specific growth phases associated with microbial growth in batch culture. If you were to add 1 ml of a pure culture of *Escherichia coli*, grown in a nutrient-rich medium, into a conical flask containing 100 ml of the same medium, then *E. coli* would grow in the S-shaped curve shown in Figure 2.4. This growth curve can be split into a series of growth phases called:

1. Lag phase;
2. Exponential (log) phase;
3. Stationary phase;
4. Death (decline) phase.

We will examine these phases of growth in more detail.

Lag phase

When cells are placed (inoculated) into fresh medium they need time to adjust; this is known as the lag phase. Some cells might be old, damaged, or lack enough ATP or ribosomes. These components must be made before the cells can start reproducing. Cells will detect their new environment and specific genes will be expressed that will enable the cells to adapt to the nutrients and conditions of their new environment. The duration of the lag phase is very much dependent on the type of change which the cells have been subjected to, they may have been placed into different media, or the temperature or pH might be different. For example, cells which have been grown in a complex, nutrient-rich medium, which are then placed into a medium containing only a single nutrient source, will have a longer lag phase. Alternatively, cells which are actively growing (exponential) and inoculated into the same medium may have a very short lag phase. It is important to note that while the cell number does not increase in lag phase, the cells are metabolically active and will be consuming some nutrients.

Figure 2.4 Time course of microbial growth curve in batch culture.

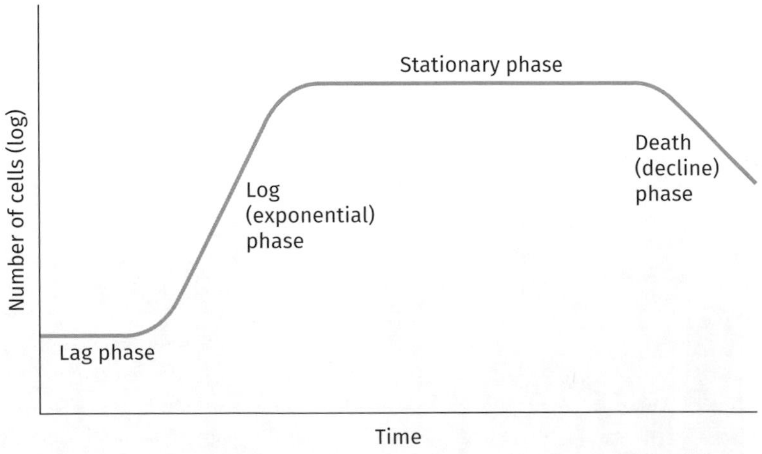

Exponential (log) phase

Once the cells have adapted to their new environment in the culture they begin to consume nutrients and synthesize cellular components such as proteins and lipids. This starts the exponential growth phase where the cells grow and divide at their specific rate (μ), causing the cell population to double over a specific length of time. This doubling time is dependent on the microbe itself, on the media composition, and other growth conditions. For example, *E. coli* grown with aeration at 37°C in nutrient-rich medium can double its population every 20 minutes.

As the cells consume nutrients in the exponential phase they generate **primary metabolites**; these are growth-related products that can either be intermediates or waste from breakdown of nutrients (e.g. ethanol, CO_2), or newly synthesized cellular components (e.g. protein, lipids). These primary metabolites are growth-linked and are only produced when cells are actively growing. In this phase you can force cells into a metabolic **downshift** or an **upshift**. Downshift is where cells are moved from a good carbon source, such as glucose to a poor carbon source such as cellulose. Upshift is the reverse of this process, the movement of cells from a poor carbon source to a good carbon source.

Stationary phase

In late exponential phase, also called the deceleration phase, the doubling rate decreases due to the reduced availability of a key nutrient, build-up of waste products, or in response to cell signals (known as quorum sensing). The cell population enters the stationary phase, where the population remains constant. The physiology of the cells changes as they express new genes to efficiently use what trace nutrients remain, and produce stress-response proteins to protect against waste products. If the cells did not change their physiology at this stage they would be much more vulnerable, as they cannot react as quickly to changes in their environment. Some microbes when in stationary phase, e.g. some species of bacteria (e.g *Bacillus* spp) will differentiate into spores, or reduce their size, minimizing the volume of the cytoplasm in relation to the volume of the nucleoid, in order to be more tolerant to environmental conditions, such as heat or osmotic pressure. Different types of microbes will reach stationary phase at different cell densities. For bacteria grown in a complex medium this will be when the cell density reaches 10^9 cfu/ml or above. For eukaryotic microbes, such as protozoa, this will be lower at 10^6 cfu/ml.

As microbes differentiate during stationary phase they no longer produce primary metabolites, instead they begin to produce molecules associated with the stationary phase known as **secondary metabolites**. These are products that are not dependent on growth, but are often associated with prolonging cell survival; these include antibiotics that kill competing bacteria.

Death phase (decline)

As no new nutrients are being provided (remember batch culture is a closed system), then the cells begin to die as waste products build up, this is the death phase. The death phase is actually a highly competitive environment because when cells die they can be cannibalized by surviving bacteria. The number of mutations in the genomes of surviving cells increases, causing subpopulations of genetically different bacteria to emerge. When antibiotics are produced during the stationary phase the rate of death can also be exponential, as the rate of cell death increases with antibiotic concentration. An understanding of the death rate is essential, particularly in the areas of food preservation and antibiotic effectiveness where the number of remaining bacteria are important.

Key points

- There are three different ways that microbes are cultured: batch culture, continuous culture, and fed-batch culture.
- The growth cycle of a microbe has four phases: a lag phase, an exponential growth phase, a stationary phase, and a death phase.
- Primary metabolites are products generated in the growth phase, while secondary metabolites are generated in the stationary phase.

2.4 Cell division in the exponential growth phase

Single microbial cells such as bacteria, yeasts (single-cell fungi), and protozoa do not really increase much in their cell size as they grow. Therefore, one way we can describe the growth of these microbes is by the increase in their population size. Single-celled organisms such as bacteria reproduce by binary fission—each cell replicates its DNA, then an equatorial septum is made that separates the cell into two daughter cells. Eukaryotic single cells, however, reproduce by mitosis, which involves the segregation of chromosome pairs. Despite these differences, the same mathematical functions can be used to describe the growth of both prokaryotic and eukaryotic single cells in batch culture.

Look at Figure 2.5. Imagine you have one cell in a volume of liquid, this cell divides through reproduction and produces an identical daughter cell,

Figure 2.5 A single cell reproduces by dividing into two daughter cells, which can then each split into a further two cells, giving four cells in total. At each generation the cell population doubles, causing exponential growth.

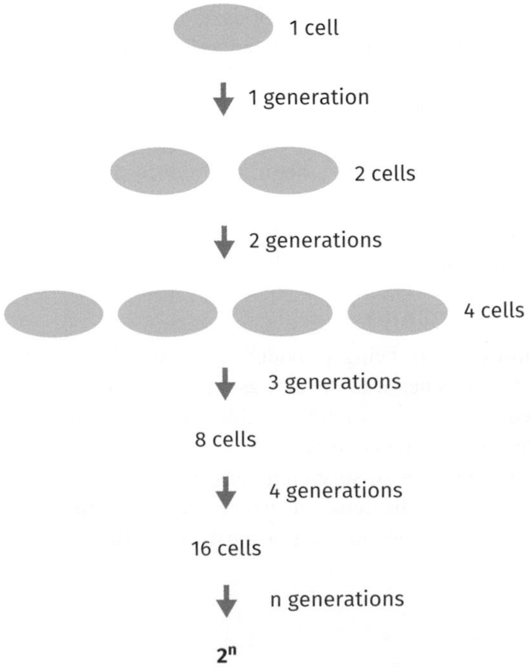

there are now two cells in the liquid. Both cells will now each produce another daughter cell, thus the population is now four cells. This is called **exponential growth**.

Because the cell population doubles with each generation, the number of cells in the population can be calculated using the formula 2^n. So to calculate the number of cells generated from a cell population greater than one, the number of cells needs to be multiplied by 2^n.

Starting with any number of cells (X_0) the number of cells after n generations will be

$$X_0 \times 2^n \quad [1]$$

Example

A single cell after five generations will produce?

$1 \text{ cell} \times 2^5 = 32 \text{ cells}$

A population of 15 cells after three generations will produce?

$15 \text{ cells} \times 2^3 = 120 \text{ cells}$

Where X_0 is the number of cells at the start of exponential growth.

During growth the number of cells of any generation can be expressed as:

$$X_n = X_0 \times 2^n \quad [2]$$

Where X_n is the number of cells after n generations starting with a population X_0.

Term in equation [1]	What it means
X_n	the number of cells after n doublings
X_0	the number of cells at the start of growth
n	the number of generation doublings

However, if we use this equation to predict the growth of bacteria you will see the trend shown plotted onto the graphs in Figure 2.6. The steps in population are caused by the bacteria doubling at each generation.

Figure 2.6 Theoretical doubling of a cell population over seven generations shown on both a linear and logarithmic scale.

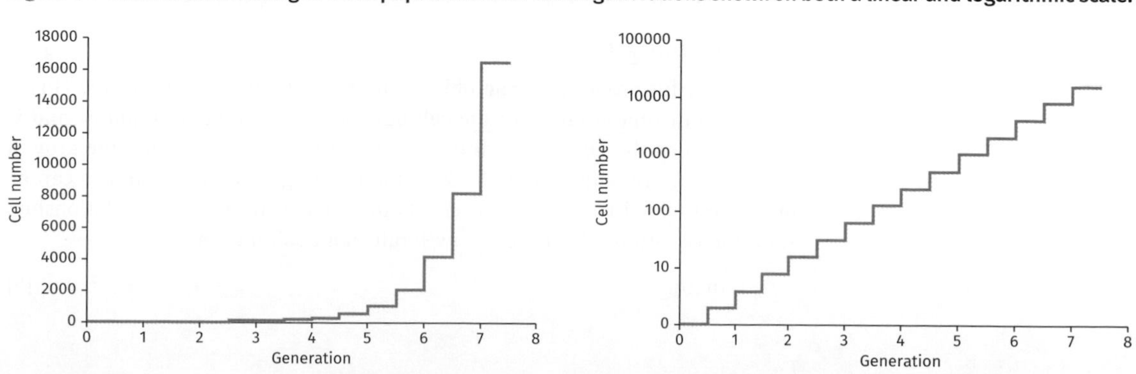

Key points

- Bacterial growth occurs through cell division.
- Each bacterial cell produces two daughter cells through binary fission causing exponential growth.

2.5 Microbial growth rates

In reality, graphs like those in Figure 2.6 do not exist. Microbial cell growth only doubles like this for a couple of generations before the steps become smooth. This is because the amount of time between each cell generation varies depending on each cell's access to nutrients, age of proteins and cell membranes, and inherited mutations. To model cell growth for a population of thousands or billions of cells we need to determine the **average rate of cell division (cell growth rate)**. This is related to the change in cell number with time using the equation [3] below.

$$X_0 \cdot e^{\mu t} = X_t \qquad [3]$$

Term in equation [3]	What it means
X_t	cell concentration after time t
X_0	cell concentration at the start of exponential growth
e	base of the natural logarithm
μ	specific growth rate (per hour)
t	time (h)

Equation [3] is commonly used in many scientific fields, including physics and biology, to describe continuous growth or decay. Most importantly this equation links the starting cell concentration, the final cell concentration, and the specific growth rate. If we take the natural logarithm of equation [3] we get the equation [4] below, which is also the equation for a straight-line graph.

$$\ln X_0 + \mu t = \ln X_t \qquad [4]$$

So, by plotting the natural log of a cell population against time, the specific growth rate of an organism can be determined. We are now going to put equation [4] into use. Take a look at the worked example in Applying the concepts 2.3.

The doubling time

Another useful property in microbial growth is the **doubling time (t_d)**. This is the length of time it takes for the cell population or biomass to double and is also known as the **generation time**. It is often more intuitive than the growth rate, as a culture with an OD of 0.2 and a doubling time of 30 minutes can be quickly estimated to reach 0.8 after 1 hour. There is an important relationship, shown in equation [5] between growth rate and doubling time.

$$t_d = \ln 2/\mu \qquad [5]$$

Applying the concepts 2.3

Example

How long will it take a culture with a starting concentration of 10^6 cell/l and a growth rate of 1.3 hr^{-1} to reach 10^9 cell/l?

$$\ln 10^6 + 1.3t = \ln 10^9$$

$$13.8 + 1.3t = 20.7$$

$$20.7 - 13.8 = 1.3t$$

$$(20.7 - 13.8)/1.3 = t$$

$$= 5.3 \text{ hours.}$$

First, the units of doubling time and growth rate are in hours (hr) and per hour (hr^{-1}), and *not* cells per hour. This is due to cells growing exponentially, i.e. the rate of growth increases with time. Secondly, the overall time for the biomass to double (t_d) is ~ 0.69 times the reciprocal of the average growth rate ($1/\mu$). If cells divided uniformly as shown in Figure 2.6 then the growth rate and doubling time would be equal, but because cells behave differently and divide at different times, the doubling time for each cell is smaller than the overall rate of growth.

Key points

- The rate of bacterial division is given by the growth rate (μ), which is related to the doubling time (t_d) by the equation
  $t_d = \ln 2/\mu$
- Bacterial growth is exponential and so the units of bacterial growth are hr^{-1} or min^{-1}.

2.6 Modelling microbial growth in batch culture

Given the ability to stay in the exponential phase, an *E. coli* cell weighing one microgram and with a doubling time of 30 minutes will produce a biomass equal to the mass of the Earth in under three days (see Applying the concepts 2.4).

Obviously this doesn't happen because the culture quickly changes from exponential phase to stationary phase. But what causes this transition and how can we model this? During the exponential phase the growth rate μ is constant, while in the stationary phase μ is zero. The transition from exponential phase to stationary phase is an intermediate known as the **deceleration phase** where μ decreases to zero. The deceleration phase can vary in length depending on the cells' interaction with waste products and the nutrients provided.

Applying the concepts 2.4

Example

Given a starting biomass of 1 pg (1×10^{-12} g) and a doubling time of 30 minutes, how long would it take a microorganism to have the mass of the moon (7.35×10^{22}kg) or the Earth (5.97×10^{24}kg)

$\mu = 0.693/0.5$ hrs (see equation [5])

$\mu = 1.39$ hr^{-1}

Using equation [4]:

$\ln 10^{-12} + 1.39t = \ln 7.35 \times 10^{25}$

$59.6 - (-27.6) = 1.39t$

= 62.7 hours to reach the mass of the moon

$\ln 10^{-12} + 1.39t = \ln 5.97 \times 10^{27}$

$64.0 - (-27.6) = 1.39t$

= 65.8 hours to reach the mass of the earth

Monod growth and the 'limiting nutrient'

A microorganism is dependent on a nutrient 'S' (substrate) for growth. When S=0 (no nutrient) there is no growth and so the growth rate μ=0. As S increases so does μ, because μ is dependent on S. However, this relationship cannot continue indefinitely. If S continues to increase, μ will eventually become constant, as something else will limit the growth rate, either the need to take in a different substrate or a limit within the stages of cell replication.

Plotting the change in μ with change in S yields the graph that you can see in Figure 2.7. The trend follows a hyperbolic curve where at low S, μ is dependent and at high S, μ is independent. There is a theoretical maximum growth rate μ_{max} and the substrate concentration where $\mu = \frac{1}{2}\mu_{max}$ is known as the Monod constant K_s

Figure 2.7 Change in growth rate of a cell with increase in nutrient (substrate).

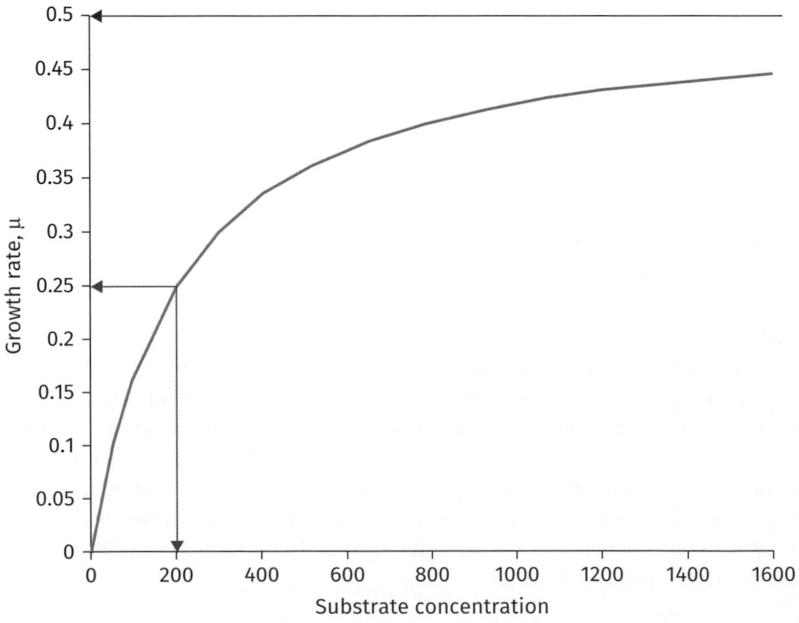

The growth rate μ shown in Figure 2.7 can be calculated at any substrate concentration S using equation [6]

$$\mu = \frac{\mu_{max} \cdot S}{K_s + S} \quad [6]$$

Term in equation [6]	What it means
μ_{max}	The rate at infinite substrate concentration and so the maximum theoretical rate
K_s	The Monod constant, a measure of affinity of an organism for a substrate

This is known as the Monod equation, and is a similar formula to the Michalis-Menton equation used in enzyme kinetics. It is an example of a 'black-box' model of bacterial growth where the thousands of different cellular reactions that occur during transport and metabolism of a substrate are generalized in a single term. This term is the Monod constant, K_s, and is a measure of the overall affinity of an organism for a particular substrate.

Knowing the Monod constant can be useful, as cells with a high K_s (low affinity) for a limiting nutrient will have a short exponential phase and a longer stationary phase, while cells with a low K_s (high affinity) will have a longer exponential phase and a shorter stationary phase. The Monod equation can also be useful for modelling the transition from deceleration into stationary phase, as the concentration of the limiting nutrient decreases during microbial growth. For an example, see Applying the concepts 2.5.

Applying the concepts 2.5

Example

A microorganism can grow on lactose with a K_s of 0.3 mM and μ_{max} of 0.9 hr^{-1}, and on maltose with a K_s of 1.4 mM and μ_{max} of 1.5 hr^{-1}. You grow the organism on a feedstock containing containing both lactose and maltose at equal concentrations of 0.9 mM. Which nutrient is most likely to be limiting?

For both nutrients work out the growth rate using equation [6].

$\mu_{lactose} = 0.9 \times 0.9 / (0.3 + 0.9)$

$= 0.675\ hr^{-1}$

$\mu_{maltose} = 0.9 \times 1.5 / (1.4 + 0.9)$

$= 0.587\ hr^{-1}$

The growth rate is highest on lactose, so this nutrient is most likely to be limiting. A more concentrated feedstock (e.g. 10 mM) would be growth limited by maltose.

The Monod equation is relatively accurate at modelling cell growth on low nutrient concentrations. At higher nutrient concentrations, growth inhibiting factors such as pH and waste products can influence growth rate and the Monod model will no longer work. Other models have been developed to better model bacterial growth, including the Contois model, where the biomass concentration (x) is included along with the substrate concentration. This equation [7] allows the modelling of inhibitory effects at high biomass concentrations and can model growth kinetics more accurately than the Monod equation in some cultures.

$$\mu = \frac{\mu_{max} \cdot S}{xK_s + S} \qquad [7]$$

In addition to the models discussed in this chapter, there are dozens of other equations used to model batch culture kinetics through the black box approach. Normally black box models are chosen on the basis that they fit the data most accurately, rather than for any clear justification involving the metabolic details of the culture process.

Modelling of product formation using the biochemical pathways of the cell is called metabolic flux analysis. This approach can only be used if the metabolic pathways of an organism are well understood and allows the user to optimize product formation by controlling a range of genetic and environmental aspects. For example, the production of butanol and acetone by *Clostridium acetobutylicum* is optimized by ensuring that butanol and acetone production is maximized and acetate and butyrate production is suppressed (Figure 2.8).

Figure 2.8 The metabolic pathway of acetone and butanol biosynthesis by *Clostridium acetobutylicum*. This model contains 17 enzymes (E1–E17) that generate 12 pathways with associated rates (r). By controlling the rates butanol production can be optimized.

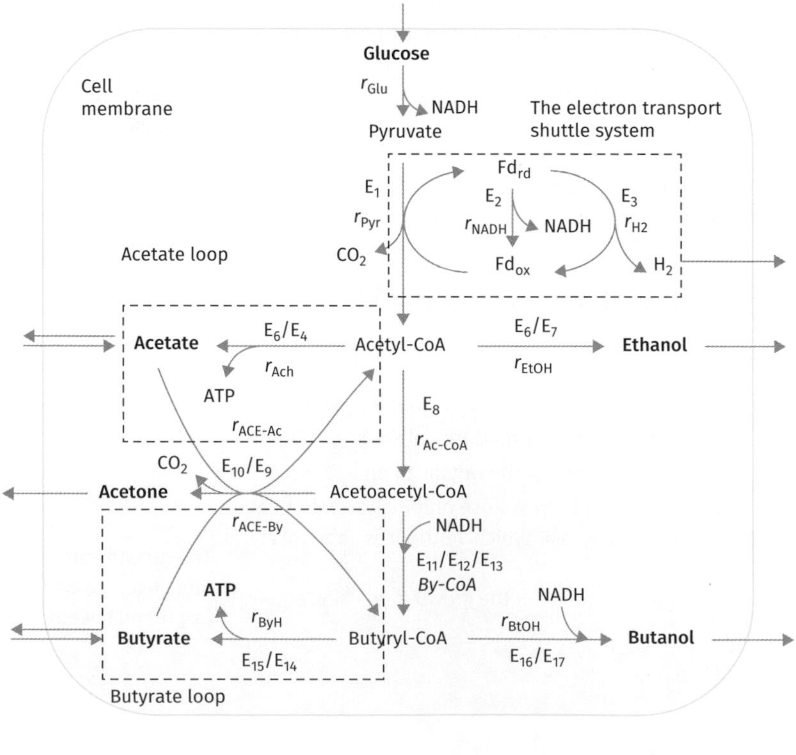

The yield coefficient and productivity in batch culture

The relationship between product formed and substrate consumed during microbial growth is key to optimizing the efficiency of a process. This relationship is usually described as a **yield**, which is the ratio of product formed over substrate consumed. It is a measure of the efficiency, or productivity of a process, rather than an absolute value of product formed. Yield can be calculated by equation [8] below.

$$\text{Yield}\left(Y_{SP}\right)=\frac{\left[\text{final product}\right]-\left[\text{initial product}\right]}{\left[\text{initial substrate}\right]-\left[\text{final substrate}\right]} \quad [8]$$

Example of unit calculation

A yeast culture ferments 100 g of sugar into 5 g of yeast and 45 g of ethanol.

The yield on ethanol is:

0.45 g ethanol g^{-1} sugar

Yield of biomass is:

0.05g yeast g^{-1} sugar

Although it is a ratio, yield values are not unitless but have units of *unit* product per *unit* of substrate. The final yield can be calculated from the final product concentration and initial substrate concentration, assuming that there is no initial product and complete substrate conversion.

It is useful to consider how the yield of a process changes during a batch culture experiment. For instance, in a batch culture run the yield will vary depending on whether the product is a primary (growth linked) or secondary (non-growth linked) metabolite. The yield of a **primary metabolite** will be constant during the exponential phase and decrease during the stationary phase, while the yield of a **secondary metabolite** will be low during the exponential phase, and then increase during the stationary phase.

It is also possible to measure the productivity, which is the rate that product is produced using the yield and initial substrate concentration.

$$r=\frac{\left(\text{product formed}\right)}{t_t} \quad [9]$$

Term in equation [9]	What it means
r	Productivity of a process
t	Total duration of a culture process

Where r = productivity and t = total time of run from inoculation to harvest. This time always includes both lag and exponential growth phases, and may also include stationary and (rarely) death phase. This productivity term is very important when it comes to the economics of a process, as costs in terms of maintenance, energy requirements, and personnel have to be factored into the cost-effectiveness of a process—the faster a product can be generated the more a process is economically viable.

Applying the concepts 2.6

Two yeast strains are grown on 100 g of sugar in separate cultures. After four days Yeast A has consumed 98 g of sugar and produced 40 g of ethanol. Yeast B has consumed 30 g of sugar and produced 13 g of ethanol. Calculate yield and productivity and decide which yeast is more suitable for ethanol production.

Yeast A

Yield = 40/98 = 40.8 g ethanol/g sugar

Productivity = 40 / 4 = 10 g / day

Yeast B

Yield = 13/30 = 43.3 g ethanol/g sugar

Productivity=13 / 4=3.25 g / day

The yields of the two yeasts are not too different, with Yeast B having a slightly higher yield than Yeast A. However, the productivity of Yeast B is very low, almost three times slower than Yeast A, indicating that Yeast A is the more efficient yeast for biotechnology purposes.

Maintenance

In order to survive cells must expend energy in cell repair and homeostasis. This means that some of the energy obtained from substrate always goes towards maintaining the cells rather than cell growth; this contribution is known as maintenance. The yield obtained is related to the growth rate by the following equation

$$Y_{SP} = \frac{\mu}{Y_{max}\ \mu + m_s} \qquad [10]$$

Term in equation [10]	What it means
Y_{max}	Theoretical maximum yield
m_s	Maintenance coefficient. How much substrate is used by the organism during homeostasis

As the growth rate increases the closer Y_{SP} gets to Y_{max}, while at low growth rates the influence of maintenance coefficient increases and Y_{SP} decreases. Thus for primary metabolic products it is advantageous to grow cultures rapidly in order to decrease the substrate that would otherwise be used for maintenance.

Key points

- In batch cultures, the transition between growth and stationary phases is called the deceleration phase.
- The growth rate is restricted by a 'limiting nutrient'. The Monod equation models the change in growth rate with limiting nutrient concentration and gives the Monod constant K_s, the affinity of the microbe for the limiting nutrient, and μ_{max}, the theoretical maximal growth rate.

- Metabolic flux analysis can be used to optimize product formation of a microbial culture but requires a good understanding of the biochemical pathways used by the organism.
- Yield and productivity are used to analyse and optimize efficiency of microbial products. For poorly growing cells the amount of nutrients consumed in maintaining the microbial population must also be factored.

2.7 Continuous culture

These are open culture systems where new fresh medium is constantly added to a culture, and an equal volume of culture is constantly removed. Because there is a continuous supply of fresh nutrients and a continuous removal of cells and waste, the stationary phase is never reached. This means that microbial cultures can be kept in exponential phase at a constant cell mass for extended periods of time so primary metabolites can be constantly produced.

Modelling growth kinetics in a continuous culture

The properties of an open culture system are dictated by the feed rate and the culture volume. These two parameters allow the calculation of the dilution rate, shown in equation [11], which is an important parameter in understanding how microbes grow in a continuous culture.

$$D = F/V \qquad [11]$$

Term in equation [11]	What it means
D	Dilution rate (units typically hr^{-1})
F	Feed rate, the volume of nutrient continuously loaded into the culture per hour
V	Volume of culture

The dilution rate is related to growth rate: the more nutrients which are received as a result of increasing flow rate, the faster the cells can replicate, which decreases generation time. Because the culture is removed as nutrients are added, the biomass in the fermenter vessel remains constant; this is known as the steady state and occurs when the dilution rate and growth rates are equal (equation [12]).

$$D = \mu \qquad [12]$$

This means that the generation time of the bacterial culture can be directly controlled by adjusting the dilution rate (see Figure 2.9). As the dilution rate and growth rate are equal in steady state, both μ_{max} and the K_s can be calculated by changing the dilution rate and measuring the nutrient concentration. Using the Monod equation from equation [7]:

$$D = \frac{\mu_{max} \cdot S}{K_s + S} \qquad [13]$$

In this form of the equation, S is the residual nutrient concentration, which can be determined from the effluent.

Figure 2.9 Effect of changing dilution rate on generation time, biomass, and nutrient concentration. The culture attains steady state when the biomass concentration in the effluent is constant.

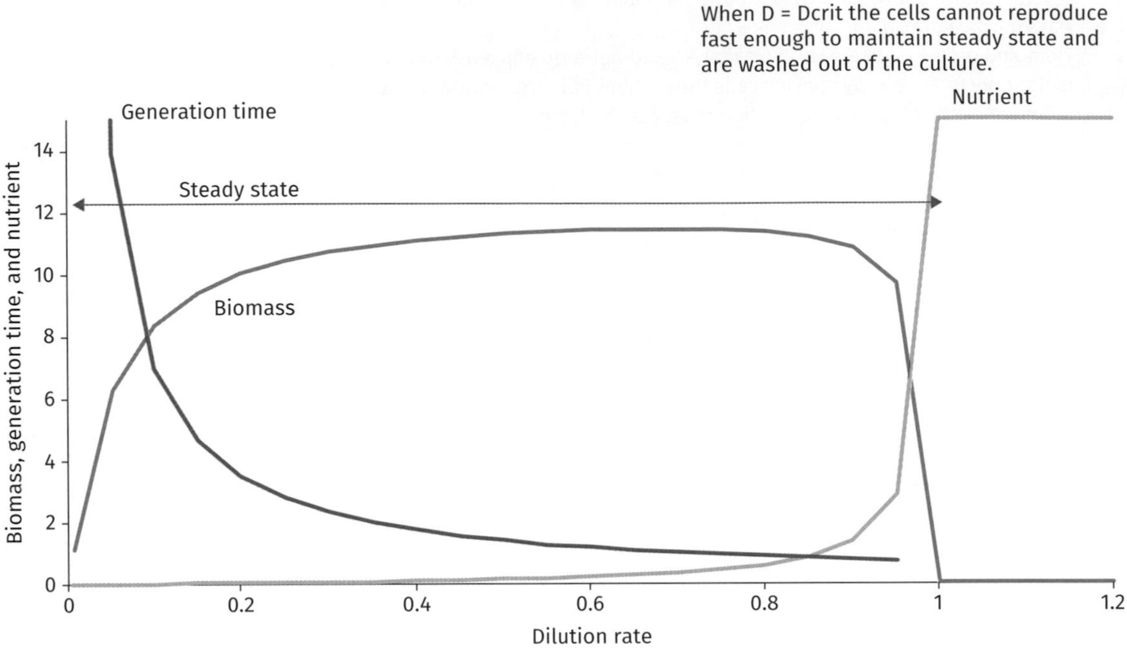

If the dilution rate is too high then the cells cannot divide fast enough to maintain a constant population and the culture is washed out of the vessel. This dilution rate is known as D_{crit} and is dependent on the maximal growth rate (μ_{max}) of the cells and the concentration of nutrient in the feed (equation [14]). At high feed concentrations D_{crit} is close to μ_{max}, but at lower concentrations D_{crit} will be much lower than μ_{max}.

$$D_{crit} = \frac{\mu_{max} \cdot S_F}{K_s + S_F} \qquad [14]$$

Term in equation [14]	What it means
D_{crit}	Critical dilution rate, the maximum dilution rate that can support the steady state culture
S_F	The concentration of substrate in the feed

Yield and productivity in continuous culture

The product yield can be determined in chemostats using the product (P_{el}) and residual substrate (S) measured from the eluent.

$$Y_{SP} = (P_{el})/(S_F - S) \qquad [15]$$

As the culture is continually replaced in the chemostat any initial product (biomass inoculum) is diluted away and so is not considered. Product would also not normally be present in the nutrient feed.

Because continuous culture is constantly producing it is worth considering the effect of productivity. This is given as product produced during exponential growth.

$$r = DP_{el} \qquad [16]$$

Comparison of the productivity rate of batch culture and continuous culture (equations [10] and [19]) show that for a primary product such as biomass, the continuous culture system will be much more efficient. This is because the duration time in batch culture includes lag phase and deceleration phases, and is why continuous culture is good for obtaining primary products. Despite the increased productivity and capability to extend the exponential phase, continuous culture fermentations are not used as frequently as batch culture. This is for several reasons:

1. Left for extended periods the culture morphology changes—bacteria in a biofilm on the sides of the vessel have an advantage over planktonic cells.
2. Mutations arise in the population, giving a heterogenous culture.
3. Increased risk of contamination as an open system.
4. Batch culture fermenters are easier to adapt to different products and projects.
5. There is a continuous production of eluent which must either be stored or continuously processed.

Despite these drawbacks continuous cultures are often used to study bacterial growth and product formation. Variation of the dilution rate of a steady-state culture allows both the maintenance coefficient and maximum yield to be determined in a single experiment using equation [17].

$$P_{el} = \frac{D}{Y_{max}D + m_s}(S_f - S_{el}) \qquad [17]$$

There are also some classic examples of continuous cultures that are used frequently in biotechnology, for instance wastewater treatment plants require the continuous processing of contaminated water.

Key points

- The dilution rate is a measure of how quickly the volume of the culture chamber is replaced and is given by D = F/V.
- When a continuous culture is in steady state the dilution rate is equal to the growth rate.
- The point at which the dilution rate exceeds the growth rate is called D_{crit} and results in the microbial culture being washed out of the chemostat.
- A chemostat can extend the growth phase enabling primary metabolites to be generated for longer.

2.8 Fed-batch culture

In fed-batch culture new nutrients in the form of fresh medium are added at specific time intervals, but waste products accumulate as culture and spent medium is not removed. This is still a closed system, but the volume in the reactor increases. Fed-batch culture is often used for the production of antibiotics and baker's yeast. By controlling the addition of substrates, the total substrate concentration can be controlled, extending exponential or stationary phases. In particular, by keeping the concentration in the reactor low, the amount of intermediate metabolites and waste products can be minimized, allowing for a more efficient conversion of substrate to product and substantially increasing the overall yield (Figure 2.10).

Modelling fed-batch cultures

Like continuous cultures, the constant addition of substrate dilutes the culture volume, but as there is no removal of culture the volume in the chemostat is constantly changing. The dilution rate for a fed-batch culture is shown below.

$$D = \frac{1}{V}\frac{dV}{dt} \qquad [18]$$

This form of the equation means that, in order to maintain a constant dilution rate, the flow rate would have to exponentially increase with time. However, it

Figure 2.10 Comparison of *Bacillus subtilis* cell growth and production of antibiotic JAA by batch and fed-batch culture.

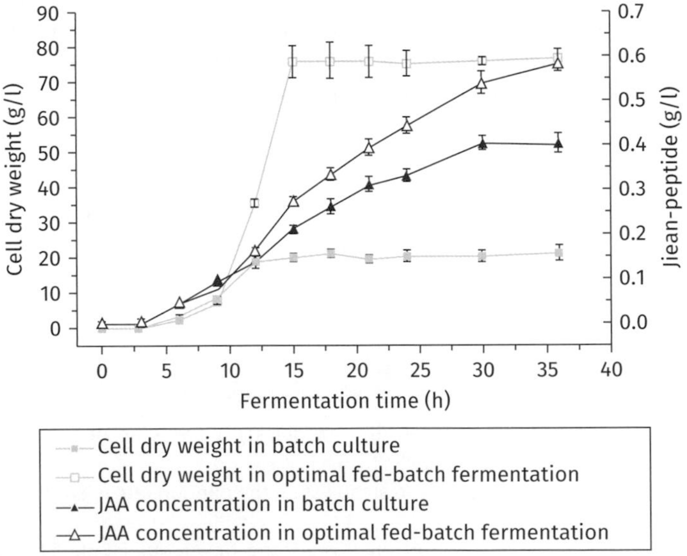

Reproduced from Juan Zhong, Xiaoyong Zhang, Yanli Ren, Jie Yang,Hong Tan, Jinyan Zhou. Optimization of *Bacillus subtilis* cell growth effecting jiean-peptide production in fed batch fermentation using central composite design. Electronic Journal of Biotechnology. May 2014 Copyright © 2014, Elsevier

is more common that the feed rate is kept constant and the dilution rate exponentially decreases. As we have seen from continuous cultures, at steady state $D = \mu$, and the same applies to fed-batch cultures.

$$\mu = \frac{1}{V}\frac{dV}{dt} = \frac{F}{V} \qquad [19]$$

Term in equation [19]	What it means
dV/dt	Change in volume over change in time
F	Feed rate, the volume of nutrient continuously loaded into the culture per hour
V	Volume of the culture at time = t

This means that, unlike batch cultures or continuous cultures, the growth rate is continuously changing. In order to minimize the effects of this the feed rate is kept constant and as low as possible so that the change in volume is minimized.

Usually, fed-batch cultures are run under conditions of very little substrate. A batch culture is inoculated and left until almost all substrate is used up and the culture has begun to enter the stationary phase. At this point the feed is turned on and substrate is added to the vessel at such a low rate that the substrate is used as quickly as it enters the reactor. In this way product accumulates with a near-complete conversion of substrate.

The yield of a fed-batch reactor is given by the equation

$$Y_{SP} = \frac{(V \cdot P) - (V_0 \cdot P_0)}{(V_0 \cdot S_0) + (F \cdot t \cdot S_F) - (V \cdot S)} \qquad [20]$$

Term in equation [20]	What it means
V_0	Starting volume of the culture
V	Volume of the culture at time = t
S_0	Initial nutrient concentration
S_F	Concentration of nutrient in the feed
S	Concentration of nutrient in the culture at time = t
P_0	Initial product concentration

By making S_F as concentrated as possible the feed rate F can be kept low, meaning that the final volume can also be kept low as well, allowing for a more concentrated product at the end of process.

A classic example of a fed-batch process is the production of baker's yeast. At high concentrations of sugar, yeast converts sugar to ethanol, which limits the amount of yeast biomass produced. Using the fed-batch process the sugar is continuously added to maintain a low concentration of sugar, which the yeast fully catabolizes to CO_2. Higher yields of yeast are generated by the complete conversion of sugar to CO_2, and the prevention of ethanol waste build-up which would then poison the cells.

Key points

- Fed-batch culture growth allows constant control of the limiting nutrient, with the result that intermediate metabolites do not accumulate.
- Fed-batch reactor processes allow for near complete conversion of substrate to product, for example in producing baker's yeast without the generation of alcohol.

Chapter Summary

- Microbial cells can be measured in a number of different ways, but all methods have their advantages and disadvantages.
- Bioreactors used for microbial growth can be operated using batch, fed-batch, or continuous culture.
- There are specific growth phases associated with microbial growth in batch culture.
- Different metabolites can be produced at different specific growth phases.
- Each bacterial cell produces two daughter cells through binary fission causing exponential growth.
- Bacterial growth can be modelled using the Monod equation.
- Yield and productivity are important parameters to determine from culture experiments.
- Continous cultures allow the exponential growth phase to be extended.
- Fed-batch cultures allow for low substrate concentrations, increasing yields, and minimizing the formation of intermediate metabolites.

Further Reading

Hoskisson, P. A and Hobbs, G. (2005). 'Continuous culture—making a comeback?' Microbiology 151: 3153–9.
A review of continuous cultures, covering the basics and many examples of how continuous cultures are used in research.

Minihane, B. J. and Brown, D. E. (1986). 'Fed-batch culture technology'. Biotechnology Advances 4: 207–18.
An older paper that looks at how fed-batch modelling can be used to control microbial metabolism.

Ratledge, C. and Kristiansen, B. (2006). Chapter 6 In: *Basic Biotechnology* Cambridge University Press.
Despite the title this is a comprehensive look at the the fundamental principles of microbial growth in bioreactors, with Chapter 6 providing derivations of all of the growth processes described here and more!

Discussion Questions

2.1 You are growing the yeast *Saccharomyces cerevisiae* in a medium which has particulate matter suspended in it. What is the best method for determining growth and why?

2.2 You want to produce an enzyme which is being produced during exponential phase, what is the best mode of operation of a reactor to ensure maximum yield?

2.3 You want to produce an antibiotic that is produced during the deceleration and stationary phases. What mode of bioreactor operation would ensure maximum yield?

2.4 The maximum theoretical yield of ethanol production from yeast is 0.511 g yeast/g glucose. How would you grow yeast to get as close to this as possible.

3

MICROBIAL BIO-PRODUCTION

Learning Objectives

- To be able to explain the role microorganisms play in bio-production;
- to be able to give an overview of the production of pharmaceuticals by microorganisms;
- to be able to describe microbial industrial enzyme production and discuss their importance in industry;
- to be able to give an overview of the production of vitamins and polymers by microorganisms;
- to be able to discuss the use of cheap substrates for microbial production of compounds for advanced biofuels;
- to be able to describe electrogenic bacteria.

Given the diversity of microorganisms it comes as no surprise that numerous microbial products are of interest to us, and that these compounds have a varied chemical nature, depending on the host that naturally produces them, and their function. Not only do we draw on homologous production—using the original producer species—but heterologous production using a genetically modified host organism has become a routine option whenever an interesting new biological molecule is discovered.

3.1 Pharmaceutical products

Microorganisms are an invaluable asset for the drug industry, each year producing diverse chemicals such as statins worth many billions of pounds. These chemicals may be produced directly by the microorganism, or by the microorganism transforming chemical compounds added to the culture medium.

Antibiotics

When we think of drugs originating from microorganisms, we probably think of antibiotics first. They have saved millions of lives since the 1940s, and you have probably taken them at some point. However, we are seeing increasing antibiotic resistance: globally, 700,000 people now die each year of infections previously curable with antibiotics. This figure could rise to 10 million by 2050 if more successful action is not taken by everyone—from members of the public to government, industry, and healthcare professionals. Biotechnology plays a crucial role there, but here we will focus on antibiotic production rather than resistance.

In 1928 Alexander Fleming was working on staphylococci bacteria when he found a mould had contaminated a Petri dish and inhibited the growth of the staphylococci. Although it was an accidental discovery, we use this principle among others when searching for new antibiotics, as you will see later in this chapter.

Antibiotics are secondary metabolites (i.e. not essential for growth or reproduction) produced close to or in the stationary phase, usually when one or several nutrients are limited (see Figure 3.1).

While primary metabolites (e.g. vitamins) are synthesized during the growth phase, secondary metabolites play other roles, for example making new nutrients accessible to the microorganism or fighting the competition (i.e. other microorganisms). The role in fighting competition is the case with antibiotics and we use this knowledge to treat infectious diseases.

You can explore some of the most commonly used antibiotics in Table 3.1.

Secondary metabolites are often complex molecules which may include ring structures, such as N-heterocycles, and their synthesis usually requires a series of enzymatic steps using substrates that originate from primary metabolites. Various basic synthesis pathways have been characterized, and are summarized in Figure 3.2.

It is important to notice from Figure 3.2 how:

- rifamycin is produced via shikimic acid;
- nystatin and chloramphenicol via chorismic acid;
- actinomycin via tryptophan;
- nocardicin via phrenic acid;
- polymyxin via phenylalanine;
- novobiocin via tyrosine.

Some antibiotics are naturally produced by more than one microbial species.

Antibiotic producers are frequently found in habitats where nutrients are scarce or their levels fluctuate, such as in the soil environment. Soil-dwelling *Streptomyces* species produce most (70 percent) of our clinically useful antibiotics.

› Biotechnology industries are businesses and processes have to be developed to maximize profits following the principles you have already learnt in Chapter 1.

› Many bioproducts related to health are often expensive and challenging to produce because of the purity and activity that is required for use in humans and animals. Remember that this usually results in compromises having to be made in terms of the approach to production that is taken.

› Interestingly, extremophile hosts can offer ways to increase productivity, which will be covered in detail in Chapter 9. Bioproducts related to food also warrant a more specific consideration and are the focus of Chapter 4.

Figure 3.1 The production of primary and secondary metabolites in comparison. (a) Alcohol (a primary metabolite) fermentation by yeast; (b) penicillin (a secondary metabolite) production by *Penicillium chrysogenum* only after mid-log phase.

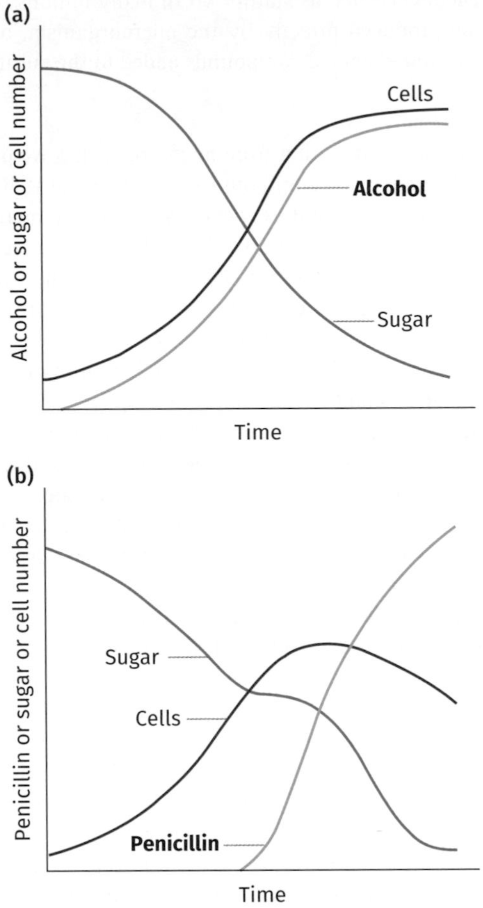

Jerzy Jankun Professor, DSc, PhD Director Urology Research Center, Editor-in-Chief Translation: UTJMS, Department of Urology, Health Science Campus, The University of Toledo, Mail Stop 1091, 3000 Arlington Ave., Toledo, OH 43614-2598, USA. Phone: 419 383 3691; FAX 419 383 3785 Email: Jerzy.Jankun@utoledo.edu

Colonies of *Streptomyces* spp can resemble fungal moulds (as can be seen in Figure 3.3) but are in fact true prokaryotes.

The search for antibiotic-producing organisms

Traditionally, the search for antibiotic producers starts with the screening of various microorganisms (e.g. from soil samples) for their inhibitory effect on others. Figure 3.4 gives an overview of the steps involved in two approaches. One approach first isolates potential antibiotic producers, and then an overlay agar is added containing an indicator organism. Antibiotic-producing isolates are identified by zones of inhibition. The second approach uses cross streaks of indicator organisms which are streaked near to a microbe which has previously been isolated. Antibiotic activity is measured by how close the indicator species will grow to the microbial isolate.

Table 3.1 Commonly used antibiotics. Examples of a range of groups are given with the producer and mode of action.

Antibiotic group	Examples	Producing organism	Mode of action
β-lactams	Penicillin G	*Penicillium chrysogenum*	Inhibit cell wall synthesis
	Methicillin Oxacillin	Semi-synthetic penicillin	
	Nocardicin	*Nocardia uniformis*	
	Ampicillin	Semi-synthetic penicillin	
Cephalosporin	Cephalothin	*Cephalosporium acremonium*	
	Cefotetan	semisynthetic	
Aminoglycosides	Gentamicin	*Micromonospora purpurea*	Inhibit protein synthesis
	Kanamycin	*Streptomyces kanamyceticus*	
	Neomycin	*Streptomyces fradiae*	
	Streptomycin	*Streptomyces griseus*	
Fluoroquinolones	Ciprofloxacin	Synthetic antibiotic	Inhibit DNA synthesis
Monobactams	Aztreonam	*Chromobacterium violaceum*	Inhibit cell wall synthesis
Carbapenems	Ertapenem	*Streptomyces* spp	
Polypeptide antibiotics	Bacitracin	*Bacillus licheniformis*	
	Actinomycin	*Streptomyces antibioticus*	Inhibit protein synthesis
Macrolides	Erythromycin	*Streptomyces erythreus*	
Other	Tetracycline	*Streptomyces aureofaciens*	
	Chloramphenicol	*Streptomyces venezuelae*	
	Lincomycin	*Streptomyces lincolnensis*	
	Cycloserine	*Streptomyces griseus*	Inhibit cell wall synthesis
	Vancomycin	*Streptomyces orientalis*	
	Polymyxin	*Paenibacillus polymyxa*	Disrupt membranes
	Nystatin	*Streptomyces noursei*	
	Rifamycin	*Amycolatopsis mediterranei*	Inhibits RNA polymerase
	Novobiocin	*Streptomyces niveus*	Inhibits gyrase
	Trimethoprim	Synthesized from gallic acid	Inhibits folic acid synthesis

Nowadays our search is not limited to the growth-based methods described in Figure 3.4; instead, we can analyse microbial genomic information to identify synthetic abilities. Yet, the development of any production process involves optimizing growth and sometimes modification of the antibiotic–as in the case of penicillins–to increase activity and therapeutic spectrum or to lessen side effects.

Production of antibiotics—using *Penicillin* as an example

A routine fermentation leads to production of natural penicillins (shaded green in Figure 3.6), but if specific precursors are added, a range of biosynthetic penicillins is produced (shaded pink in Figure 3.6). In contrast, semi-synthetic

❯ A discovered antibiotic and the producer are studied further and the production process is developed as described in Chapter 1. Carbon sources for cultivation can be lactose and glucose, or cheaper substrates such as molasses or soy meal. Complex substrates or ammonia salts provide nitrogen. Trace elements such as phosphorus, sulphur, magnesium, zinc, iron, and copper are added to the culture medium as salts.

❯ Antibiotic production is linked to nutrient limitation and thus the fermentation regime needs careful regulation. For penicillin production, corn steep liquor has traditionally been the major ingredient and acts as the nitrogen source. Lactose is used in fed-batch culture (remember Chapter 2) as the initial carbon source to not repress penicillin formation, but then an optimized glucose feeding approach as illustrated in Figure 3.5. is required for optimized penicillin production. The measurable glucose in the medium needs to be kept to a minimum to create the necessary stationary phase conditions.

Figure 3.2 Aromatic antibiotics. This flow diagram summarizes pathways from various microorganisms, but no single one can produce all these compounds. The key indicates how primary and secondary pathways are involved.

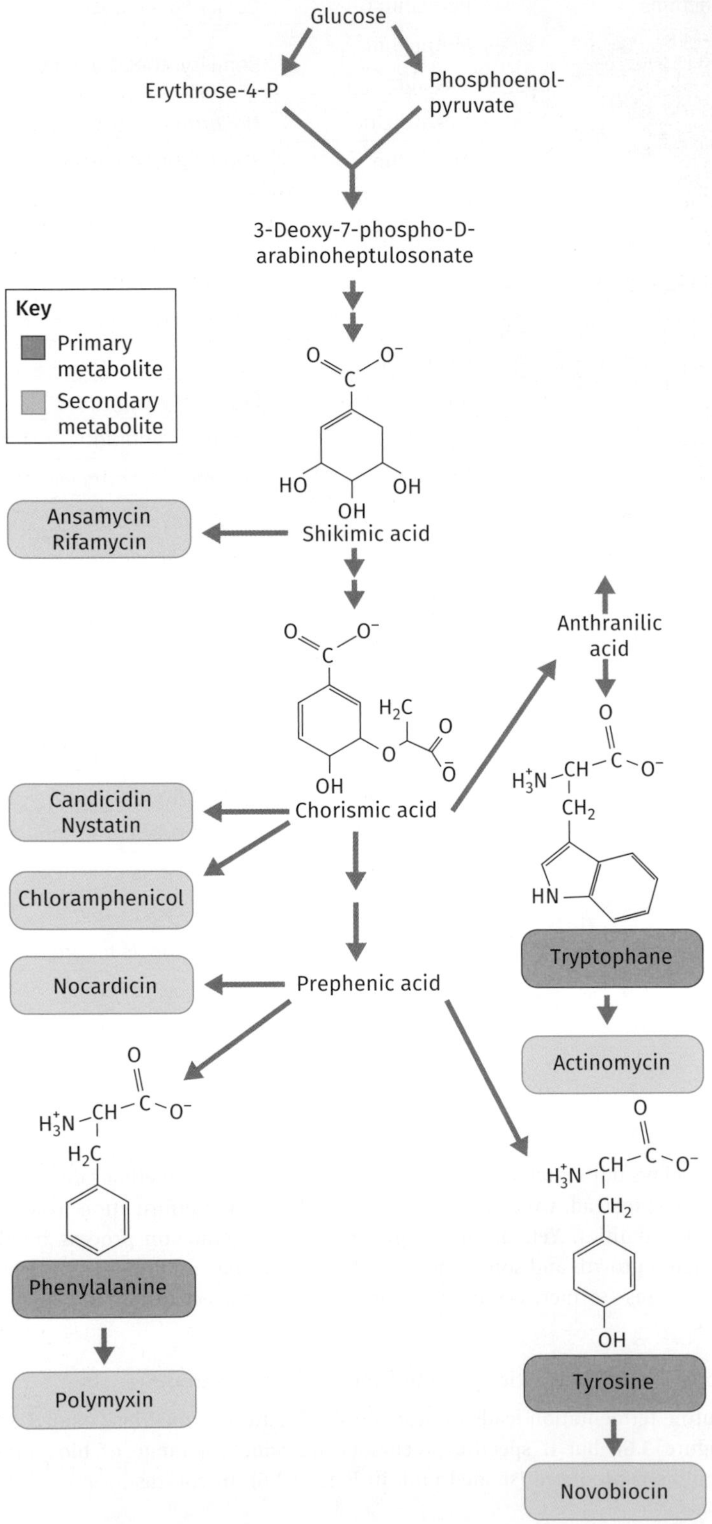

Figure 3.3 *Streptomyces coelicolor* colonies. The fluffy appearance indicates sporulation.

Matt Hutchings, University of East Anglia.

penicillins (shaded purple in Figure 3.6) are produced by chemically adding side-chains to the core structure in order to increase the activity spectrum.

Figure 3.7 gives an overview of the purification process. Penicillin has been secreted by the host cells into the medium, which is readily harvested following separation from the biomass. The pH is then lowered and solvent extraction frequently follows. Penicillin is then concentrated and crystallized.

The bigger picture panel 3.1
Bioprospecting and biopiracy

For centuries explorers and scientists have visited remote areas and discovered new plant and animal species, and brought these back as trophies for display and subjects of research. More recently this included micro-organisms (e.g. from soil samples) that produce bioactive compounds for commercialization. Little or no compensation was given to the actual owners of the resources or of the knowledge, which were often developing countries or indigenous people. Since 1993 the Convention on Biological Diversity has secured rights for these countries to control access to their natural resources. However, the convention has not been ratified by all countries, including the USA, and local laws are not fully in place everywhere. According to the Convention, bioprospecting contracts grant consent to access resources and identify the sharing of benefits. The contracts are not always fair, thus biopiracy still exists. What are the issues in this complex situation, and are there any that have not yet been alluded to? What are the challenges in addressing any of the issues?

Figure 3.4 Isolation and screening of antibiotic producers. (a) Selective media are used to isolate e.g. *Streptomyces* spp., and one indicator organism (here *Staphylococcus aureus*) embedded in the overlay is used to screen for antibiotic producers via zones of growth inhibition around the producer colony. (b) The cross streak method uses a range of indicator organisms (here *Escherichia coli, Bacillus subtilis, Staphylococcus aureus, Klebsiella pneumoniae, Mycobacterium smegmatis*) streaked near a producer colony to screen for inhibition and thus the activity spectrum of the antibiotic.

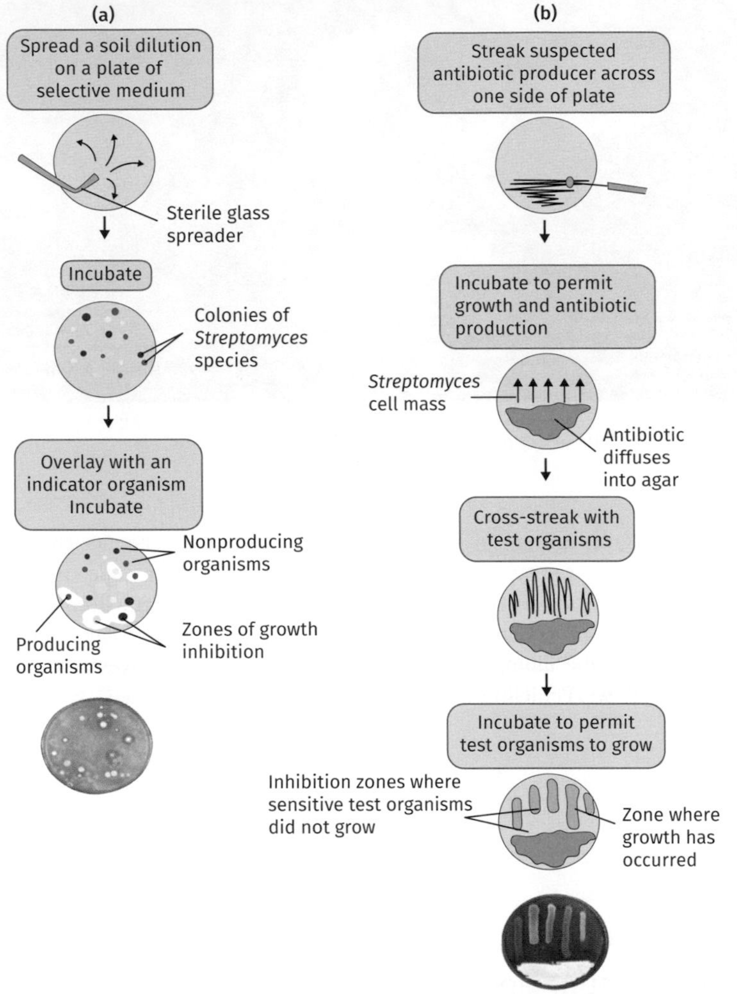

(a) Science Source / Science Photo Library (b) © SCIENCE SOURCE / SCIENCE PHOTO LIBRARY

Key points

- Microorganisms produce compounds worth many billions of pounds each year, directly or by transforming chemical compounds.
- Antibiotics are secondary metabolites produced close to or in the stationary phase, usually when nutrients are limited, requiring a carefully regulated fermentation regime.
- Producers such as *Streptomyces* spp. are for instance found in soil.
- Screening for antibiotic producers can be based on isolated microorganisms or culture-independent methods using isolated DNA.

Scientific approach panel 3.1
Culture-independent discovery of natural products

Research of natural microbial ecosystems was limited by our inability to cultivate more than 99 percent of known microorganisms. Culture-independent, or **metagenomic**, methods extract and analyse DNA (e.g. screen for known patterns of encoded metabolic pathways) from the environment (eDNA), including extreme habitats, followed by heterologous expression of the DNA to identify certain synthetic abilities based on the same principles used for cultivated isolated potential producers. There are two approaches with steps as follows:

- Metagenomic data extraction
    - Construction of a metagenomic library
    - Screening of known sequence or function

OR

- Biosynthetic profiling of metagenomes based on the sequence
    - *In silico* analysis for novel sequence markers
    - Construction of metagenomic libraries of these markers
- Recovery and sequencing of any novel gene clusters;
- Activation of any **silent clusters**;
- Expression, isolation, and characterization of any novel products.

These approaches are not only relevant to antibiotic discovery, but to other therapeutic agents, towards **bioremediation** and for novel foods.

Do you think they will fully replace culture? Are there any limitations?

Figure 3.5 Kinetics of penicillin fermentation with *Penicillium chrysogenum*. Production is linked to the stationary phase.

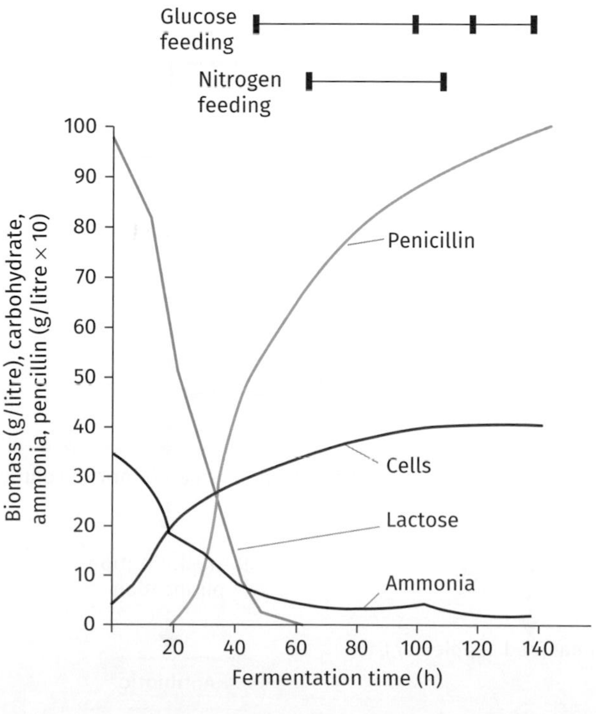

Figure 3.6 Industrial production of penicillins. The use of precursors and chemical modification leads to a range of products.

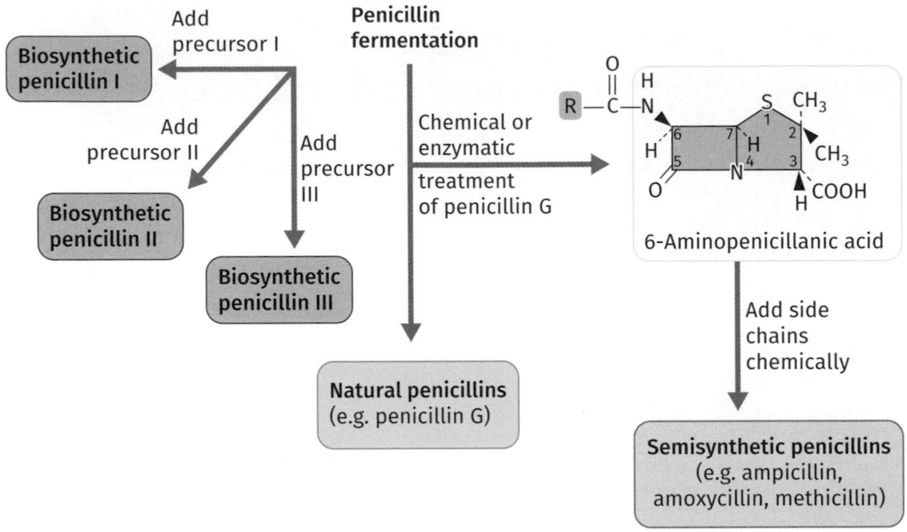

Figure 3.7 Purification of an antibiotic. The flowchart shows all necessary steps.

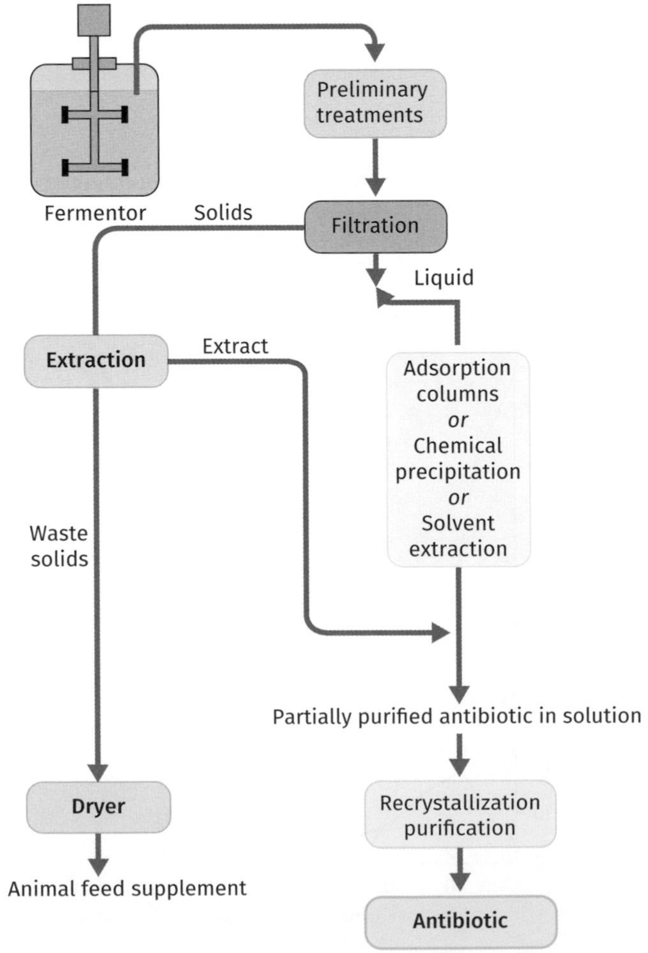

Steroids

When we hear 'steroids' we often think of anabolic steroids used to boost athletic performance to the detriment of the health of the individual. However, steroids also include therapeutic agents such as corticosteroids, used to treat asthma patients, and steroid hormones, used as oral contraceptives. These substances used to be extracted directly from adrenal glands of animals, but this practice is not only expensive, but unsafe. Much cheaper than extracting from animals or chemically synthesizing from scratch is to use microbial production such as microbial biotransformation of natural steroids (e.g. diosgenin from yams and further transformation to cortisone). Microbes which do this biotransformation include the filamentous fungi *Rhizopus nigricans and Rhizopus arrhizus.*

Some microorganisms introduce hydroxyl groups at any of the carbon atoms in the steroid molecule. C11-hydroxylation of progesterone enhances cortisol activity 3–5 times. *Corynebacterium simplex* is used to dehydrogenate C1. The resulting product is called prednisolone (see Figure 3.8); its anti-inflammatory properties make it useful in the treatment of allergies, skin diseases such as eczema, some cancers, auto-immune conditions, and to prevent transplant organ rejection.

Figure 3.8 Steroid biotransformation examples. The steps indicate the use of specific microorganisms or chemical reactions.

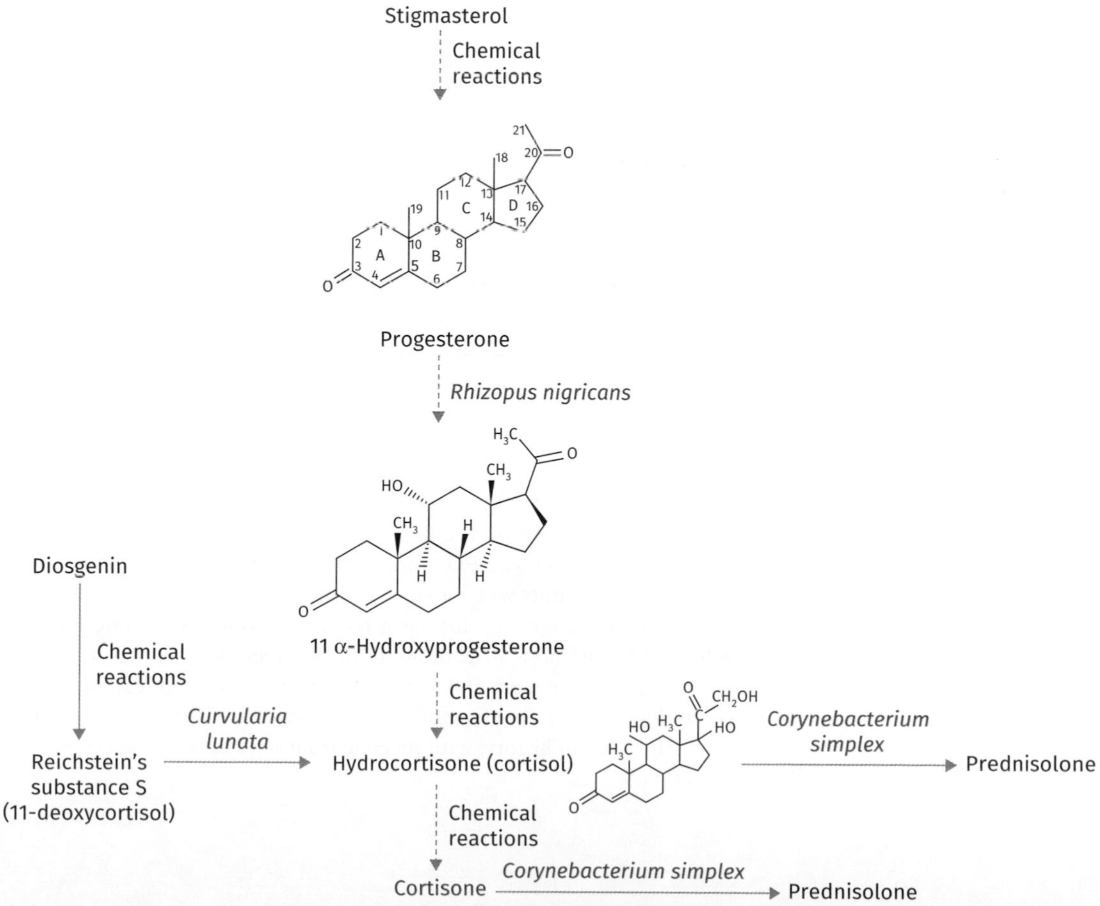

Figure 3.9 Chemical structure of cholesterol and a labelled steroid skeleton. The typical cholesterol structure with the 3-β-hydroxyl group and 5,6-double bond is indicated.

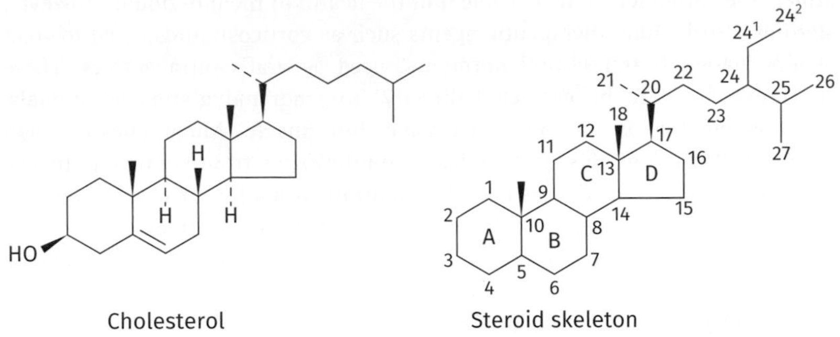

Aspergillus brasiliensis transforms progesterone into three different hydroxylated compounds: 11α-hydroxyprogesterone (as depicted in Figure 3.8), 14α-hydroxyprogesterone, and 21-hydroxyprogesterone. The thermophilic filamentous fungus *Myceliophthora thermophilas* cleaves the side chain of progesterone and also acetylates and oxidizes the 17β-alcohol of testosterone, as well as several other steroids.

Phytosterols are structurally very similar to cholesterol (see Figure 3.9) with its typical A-ring structure, the 3-β-hydroxyl group, and 5,6-double bond.

These compounds can lower cholesterol absorption in the intestine and thus reduce serum levels of low-density lipoproteins (see also Hypocholesterolaemic drugs in the next section), which in turn reduces the risk of atherosclerosis. Microbial transformation produces steroid intermediates from soybean phytosterols. The low water solubility of phytosterols, as well as the toxicity of the products to the microbial cells, is a challenge. Removing the toxic steroid products from the culture medium offers a solution. This can be done *in situ* by capturing the steroids in polymers or in the organic phase of an organic-aqueous two-phase system. Alternatively, more steroid-tolerant mutants can be selected for, or metabolic engineering can allow over expression of the genes that encode rate-limiting reactions to increase yields. Suitable bacterial hosts are, e.g., mycobacteria. Their rigid envelope acts as barrier to toxic products, and multidrug efflux pumps remove the products from the cell.

Hypocholesterolaemic drugs

Cholesterol is the main sterol in the cell membranes of animals to confer stability. Around 30 percent of the cholesterol in our bodies comes from our diet; the rest is synthesized by the liver. High levels of cholesterol can cause atherosclerosis, which can lead to coronary heart disease through the build up of plaque. Statins are drugs which reduce cholesterol by inhibiting the enzyme 3-hydroxy-3-methylglutaryl-coenzyme A reductase, which is the rate-limiting enzyme for the biosynthesis of cholesterol in the liver. Natural statins can be obtained from different genera and species of filamentous fungi (e.g. lovastatin from *Aspergillus terreus*). One type of statin, atorvastatin, is the most commercially successful drug in history, with revenue from sales already exceeding 120 billion US dollars.

Immunosuppressants

Immunosuppressants have been highly successful in preventing rejection of transplanted organs. Ciclosporin A was first discovered in the filamentous fungus *Tolypocladium inflatum* (as shown in Figure 3.10). Two other products are on the market, rapamycin and tacrolimus, and both come from actinomycetes (bacteria). All of these drugs work by binding to an immunophilin protein (an intracellular protein with an important role in the immune system). The complex which is produced interferes with the activation of lymphocytes and thus prevents the lymphocytes from eliminating the foreign antigen.

Key points

- Microbial biotransformation of natural steroids is used to produce steroid drugs.
- Statins and immunosuppressants can be obtained from filamentous fungi.

Therapeutic proteins

We have come a long way within just four decades of recombinant protein production (see Table 3.2). Recently, more than 400 marketed recombinant products were on record and more than 1,300 products were undergoing clinical trials.

In addition to reducing human suffering and saving lives, the production of recombinant proteins can result in profits in the billions.

Over time we have seen the development of new production systems, enhancing functionality based on correct protein folding and post-translational modifications, while keeping the costs low. New host systems have been introduced over time; for example, *Bacillus megaterium* and *Lactococcus lactis* have been optimized for secretion, while the filamentous fungus *Trichoderma reesei*,

Figure 3.10 Scanning electron micrograph of *Tolypocladium inflatum*. The hyphae with swollen phialides carrying conidia are shown.

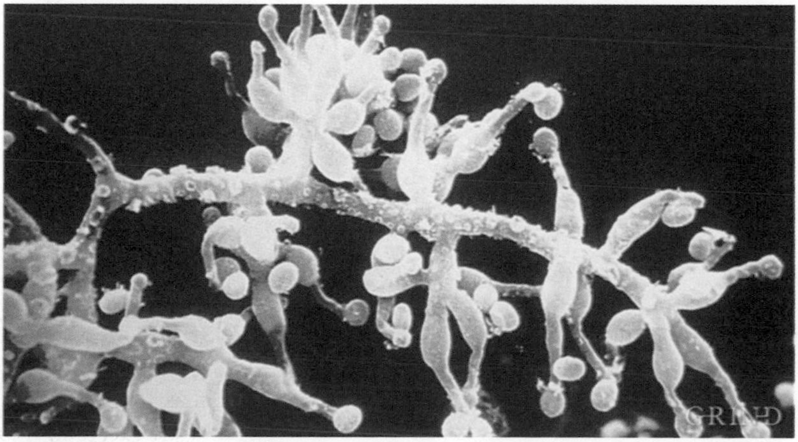

Table 3.2 Well-known milestones in therapeutic protein production. Within only a few years major bioproducts had been developed.

Year	Product	Recombinant host	Application
1982	Humulin	*E. coli*	Diabetes
1985	Protropin	*E. coli*	Human growth hormone deficiency
1986	IntronA	*E. coli*	Cancer, genital warts, and hepatitis
1986	RecombivaxHB	*S. cerevisiae*	Hepatitis B vaccine
1987	Humatrope	*E. coli*	Human growth hormone deficiency

the moss *Physcomitrella patens* or the protozoan *Leishmania tarentolae* have been optimized for their fit of **post-translational glycosylation** patterns. Endotoxin-free strains of *Escherichia coli* offer improved biosafety for low costs.

But let us go back in history for an iconic example: Humulin (shown in Figure 3.11) was the first marketable product created through recombinant DNA technology.

Before Humulin, insulin for the treatment of human diabetes had been isolated from pig and cow pancreata, but this caused immune reactions eventually neutralizing its activity

Mature human insulin decreases blood glucose concentration by signalling to some cells of the body (e.g. in the liver) to take up glucose. It is a heterodimer (51 amino acids, 5.808 kDa) comprised of an A chain and a B chain linked by two interchain disulphide bonds. Chain A also has an intrachain disulphide bond. The structure is illustrated in Figure 3.12.

Figure 3.11 Marketed insulin. The package and vial are shown.

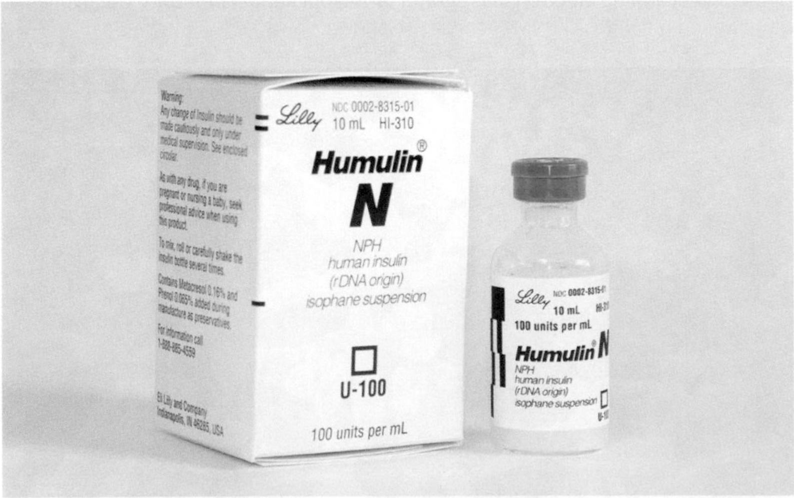

Eric Carr / Alamy Stock Photo

Figure 3.12 Structure of insulin. (a) Schematic representation of chains A and B with disulphide bridges; (b) 3D structure of human insulin with the A chain (green) covalently bound via disulphide bonds (purple) to the B chain (blue).

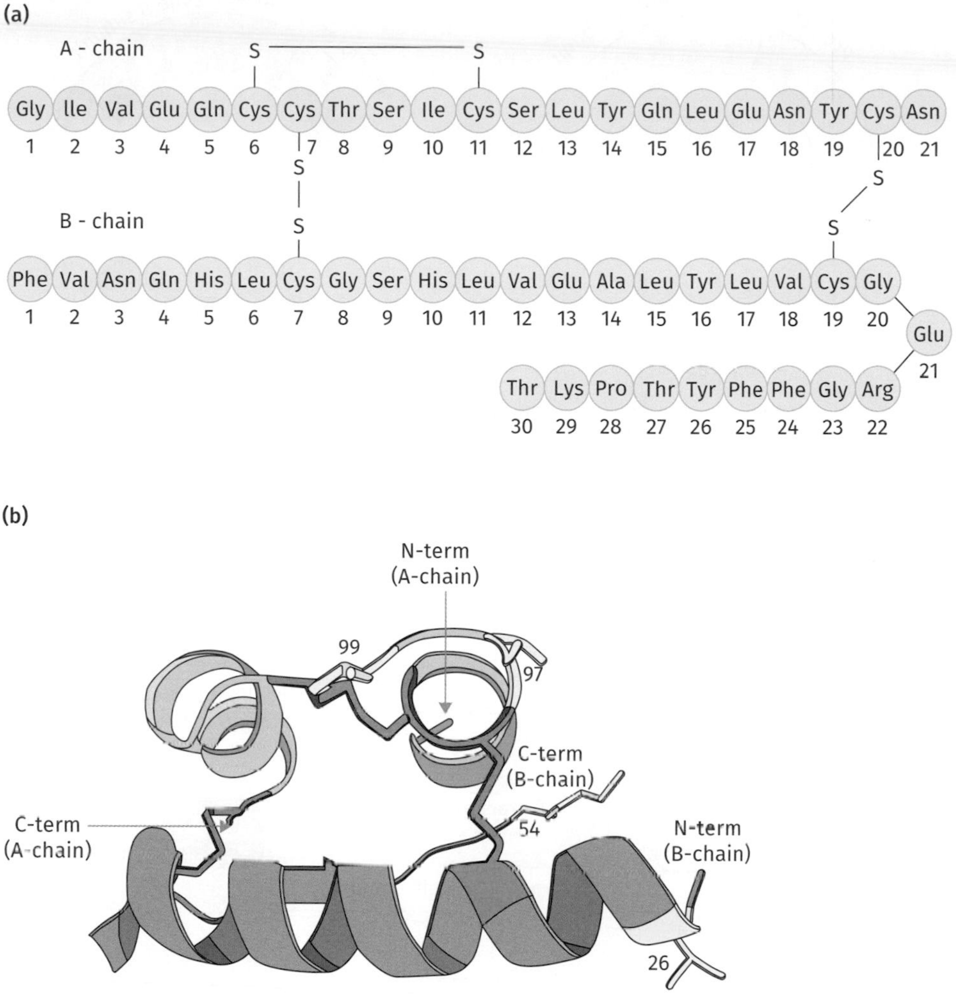

Recombinant production uses either the two-chain method (producing A and B separately; see Figure 3.13) or the proinsulin method (producing the insulin precursor—A and B chains linked together by the C-peptide). The DNA was chemically synthesized and inserted into a plasmid vector used to transform competent *E. coli* cells.

A very cost-effective approach is to express the insulin precursor in *E. coli* as inclusion bodies, which are then solubilized and refolded, resulting in the combination of both chains and the formation of the correct disulphide bonds. More currently, insulin is produced in yeasts such as *Saccharomyces cerevisiae*, secreting soluble insulin precursors into the medium. Some production approaches are compared in their production parameters in Table 3.3.

Figure 3.13 Overview of the two-chain method for insulation production. The steps involving two hosts are depicted.

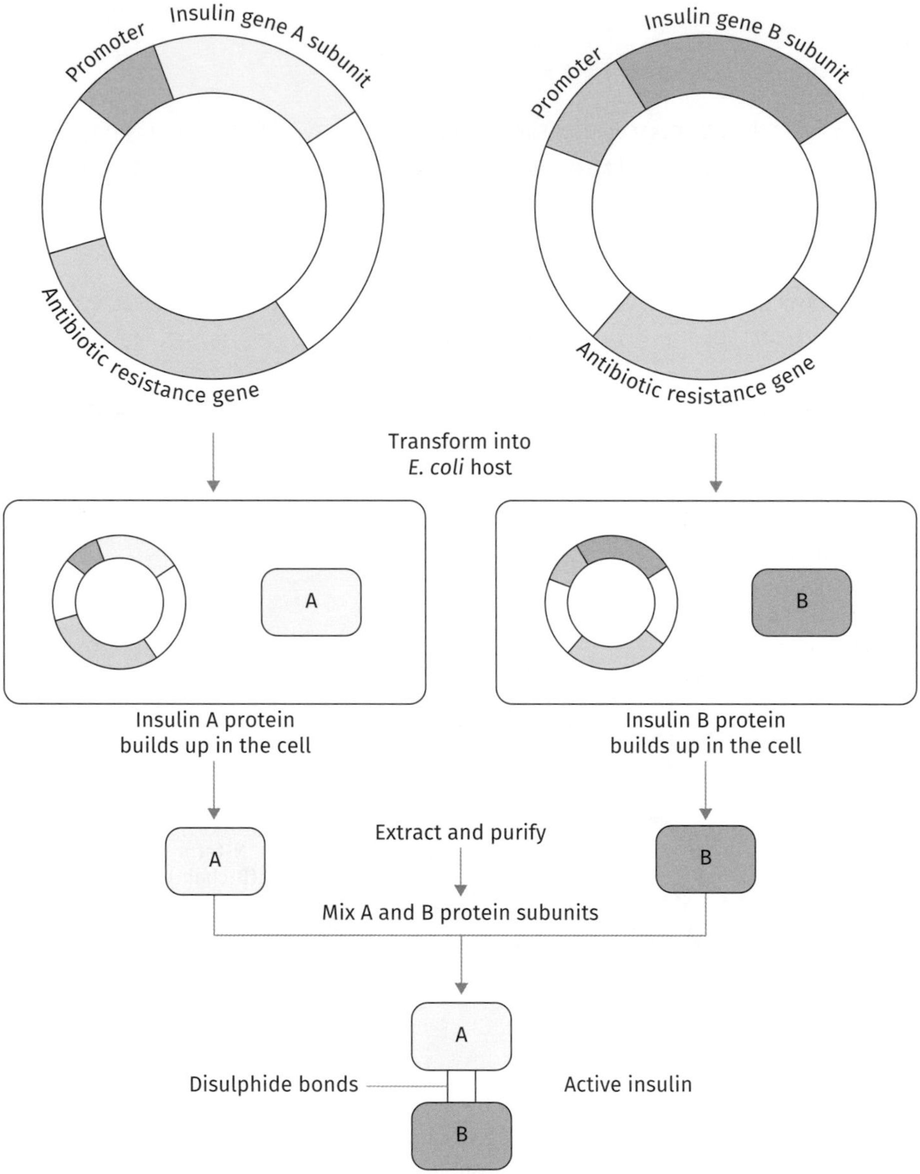

Recombinant production also allows us to modify the structure of insulin to achieve faster absorption or faster and shorter action. The first short-acting insulin analogue was Lispro in 1996. Glargine became available in 2000, where asparagine (A21) was replaced with glycine, and two extra arginines were inserted (B30). Given these changes, soluble formulation is possible at pH 4.0, but physiological pH at the injection site causes microprecipitates to form, leading to prolonged absorption. Detemir (released in 2005) has a 14-carbon fatty acid chain attached to lysine (B29), which slows its absorption.

Table 3.3 Comparison of human insulin production systems (reproduced from Porro *et al.* 2011). The systems have varied productivities which, once determined, allow a choice of production process.

Source	***E. coli***	***E. coli***	***S. cerevisiae***	***P. pastoris***
Destination of product	Cytoplasm	Secreted	Secreted	Secreted
Biomass cell dry weight (g/l)	80, in bioreactor with fed-batch culture	1.2, in shake flask with batch culture	5, in shake flask with batch culture	59, in bioreactor with fed-batch culture
Typical spec. growth rate (1/h)	0.008–0.12	Not specified	<0.33	<0.33
Typical spec. production rate (mg/gh)	14.2	3.4	0.21	0.375
Product concentration (g/l)	4.34	0.009	0.075	3.075
Productivity (mg/lh)	1,085	4.01	1.04	17

The World Health Organization estimates that over the next 10–20 years insulin sales will grow five-fold. Yet current technology is unable to meet this demand because production capacities are limited and costs are too high. Further, alternative experimental ways to administer insulin (e.g. inhalation, oral uptake) require higher doses and a higher global insulin production.

In response to this challenge, scientists have already been working on innovative approaches. Recombinant human insulin can be produced in oilseeds of the plant *Arabidopsis thaliana*. Lettuce leaves can accumulate stable proinsulin to levels that exceed half of total leaf protein, successfully fed to mice. Similar experiments have been conducted with tobacco, and it has been calculated that 20 million daily insulin doses can be produced in a tobacco plantation roughly the size of a football field in one year.

Case study 3.1
Insulin production in transgenic plants

Transgenic plants are a cost-effective host without concerns regarding pathogenicity and post-translational modifications follow eukaryotic patterns similar to that in humans. Human insulin has been successfully produced to a high level in the oilseeds of the plant *Arabidopsis thaliana*. The oilseeds have storage organelles known as oilbodies with outer wall proteins called oleosins. Recombinant proteins fused to oleosin can be targeted to these oilbodies. Oilbodies can be readily separated from other cell components using liquid-liquid phase separation, and, following cleavage from oleosin, the insulin can be easily recovered and matured for biological activity. The insulin does not need to be isolated immediately but can be stored in the oilseeds as long as necessary.

Discuss potential limitations. Could these be resolved?

Key points

- Production of therapeutic proteins has seen considerable advancement since the initial creation of recombinant vectors.
- Novel microbial host systems have been developed, as well as sophisticated production processes.
- Product tailoring allows for improved therapeutic activity with fewer side effects and optimized shelf life.

Vaccines

Vaccines including recombinant vaccines (see Table 3.4) are a global success in preventing disease. Moreover, use of vaccines crucially reduces the need for treatment with antibiotics and thus limits antibiotic resistance.

RecombivaxHB against Hepatitis B, produced in the yeast *Saccharomyces cerevisiae*, has already been mentioned as a historical milestone. The Hepatitis B virus is a small enveloped DNA virus transmitted via blood or other body fluids. More than 2 billion people are infected worldwide, with more than 350 million at risk of dying prematurely due to hepatitis, liver cirrhosis, and liver cancer.

In 1978 William Rutter's lab cloned an antigen of the Hepatitis B virus, and in 1981 a yeast expression system to produce it was published. The US Food and Drug Administration granted a licence for this first recombinant vaccine, and in 1987 RecombivaxHB became available.

Assembly of the antigene is not successful in most eukaryotes, but it is in *Saccharomyces cerevisiae*. The fermentation medium contains yeast extract, soy peptone, amino acids, glucose, and mineral salts. Self-assembly of the protein and secretion into the medium occurs. Purified protein is formaldehyde-treated in phosphate buffer and subsequently co-precipitated with potassium aluminium sulphate to produce the bulk vaccine with the adjuvants aluminium hydroxyphosphate sulphate. The year 2017 saw a global production shortage, which makes alternative production and delivery systems more urgent. Plant expression systems found in corn, lettuce, and potato are already being researched for oral delivery.

Table 3.4 Currently available recombinant vaccines. The product names are listed and a characterization provided.

Product	Characterization
RecombivaxHB	Against Hepatitis B
Engerix-B	Against Hepatitis B
Gardasil	Quadrilent against Human Papillomavirus (Types 6, 11, 16, 18)
Gardasil 9	9-valent against Human Papillomavirus
Cervarix	Bivalent against Human Papillomavirus (Types 16, 18)
Shingrix	Adjuvanted against Herpes Zoster

Key points

- Recombinant vaccines are crucial in preventing infectious diseases globally, thus limiting antibiotic resistance.
- RecombivaxHB, produced in the yeast *Saccharomyces cerevisiae*, was the first recombinant vaccine to be produced and marketed.

3.2 Fine chemicals

Unlike pharmaceutical products, fine chemicals do not usually save lives but make our lives more comfortable or pleasurable. A number of fine chemicals will be covered in this section including enzymes, vitamins, and citric acid.

Enzymes

Microorganisms produce and secrete exoenzymes into their environment, to access and break down polymeric nutrients into subunits as substrates for their metabolism. These enzyme activities are also useful for us in various contexts.

> We will explore in Chapter 4 how microbial enzymes are used in relation to food—for tenderizing meat, preparation of fruit juice, or improving digestibility of animal feed.

The most important non-food-related application for microbial enzymes is in the laundry industry.

The global market for industrial enzymes is estimated to reach about 6 billion US dollars by 2020. The detergent industry is the largest enzyme-using sector. Much of the dirt in our clothes is easily removed if broken down and made more water soluble. This is done by so-called 'biological washing detergents', allowing the use of less detergent, at lower temperatures, and with fewers chemicals persisting in the environment; altogether, making laundry more environmentally friendly and sustainable. Such enzymes also feature in dishwasher detergents and are used in waste management.

How do they work? Amylases break down starch into sugar. Proteases do the same with protein stains, breaking them down into polypeptides and amino acids. Lipases deal with fats, hydrolysing them into fatty acids and glycerol. Cellulases, modifying the cellulose contained in cotton fabrics, brighten the colour of fabrics by degrading the microfibrils formed in the cellulose due to wear and tear of the clothes. Fewer microfibrils also make the fabric feel softer.

Such exoenzymes are frequently produced by soil organisms:

> Moreover, protein engineering is performed to increase basic enzyme activity or stability, as you have already learnt in Chapter 1.

- proteases from *Aspergillus* sp. and *Bacillus* sp.;
- amylases from *Bacillus licheniformis* and *Aspergillus oryzae*;
- cellulases from *Trichoderma reesei*, *Bacillus* sp., and *Clostridium* sp.;
- lipases from *Chromobacterium* sp. and *Candida antarctica*.

Many laundry enzymes are also produced in genetically modified microorganisms to enhance yield or to lower costs.

The amount of enzyme used in detergents is fairly small, but they must be stable enough to remain active under harsh conditions in the washing cycle or in concentrated liquid or dry powder detergents (which include components such as surfactants, oxidants, and feature a pH as high as 10). The lipase of e.g. *Acinetobacter radioresistens* is optimally active at pH 10 and stable between pH 6 and 10.

Such so-called extremozymes are isolated from alkaliphiles and other extremophiles. You may already be familiar with *Taq* polymerase or *Pfu* polymerase in polymerase chain reactions (PCR), which follows the same thinking and can be used at high temperatures. We will discuss PCR in Chapter 9. Cold-active and halostable enzymes also occur naturally.

Vitamins

It is estimated that half the population take vitamin supplements on a daily basis. Vitamins used to be made from plant and animal biomass, but now microorganisms are used as the major source. Vitamins such as thiamine, pyridoxine, pantothenic acid, β-carotene (pro-vitamin A), and ergosterol (pro-vitamin D) are examples of the successful commercial production of microbial primary metabolites produced during optimal growth.

Biotin, also known as vitamin B7 or H, is a water-soluble cofactor of carboxylases. It strengthens hair and nails and is used as a nutritional supplement and in cosmetic products. The large demand could be met by chemical production (Figure 3.14 shows the chemical structure), but production in microorganisms (e.g. fungi and *Streptomyces* sp.) is more environmentally sustainable.

β-carotene is an antioxidant and forms retinol (vitamin A) for vision. It is also used in cosmetic and pharmaceutical products, as a food colouring agent, and animal feed supplement. The long isoprenoid backbone is characteristic of its chemical structure (see Figure 3.15).

β-carotene is produced by the filamentous fungus *Blakeslea trispora*. This production can be enhanced by random mutagenesis After fermentation the medium is filtered off and solvent extraction of the dried biomass takes place, with subsequent crystallization of pure β-carotene. *Escherichia coli* is used for recombinant carotenoid production.

❯ This links our discussion to food-related applications in Chapter 4, where you will also read about riboflavin (from *Ashbya gossypii*), vitamin B12 (from *Propionibacterium shermanii* and *Pseudomonas denitrificans*), and ascorbic acid (from *Acetobacter suboxydans*).

Ergosterol is also called pro-vitamin D2. In humans it is converted into ergocalciferol (vitamin D2) by UV light from the sun, and is crucial for bone development. Recombinant ergosterol production utilizes *Saccharomyces cerevisiae* grown on cheap cane molasses in fed-batch fermentation. If the yeast is exposed to UV light, conversion into vitamin D2 occurs. If the exposure follows controlled protocols, the yeast remains viable and can still produce carbon dioxide, and so vitamin-enriched bread and other baked products can be manufactured.

Citric acid

While we associate citric acid with citrus fruits, it is a metabolite of plants and animals. It is safe and biodegradable, and has a broad range of uses in food and beverages, detergents, pharmaceuticals, cosmetics, and toiletries. In

Figure 3.14 Chemical structure of biotin.

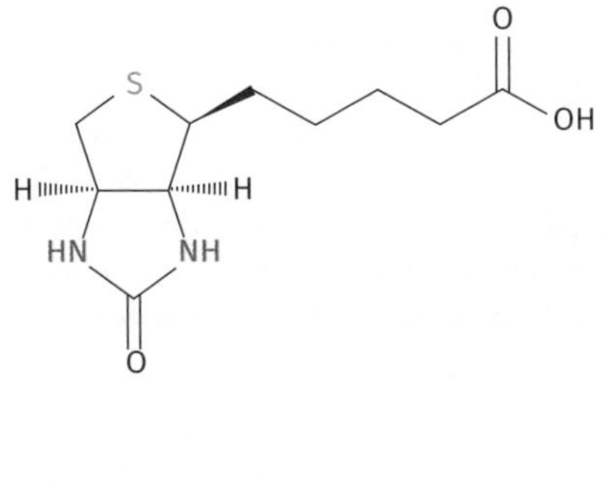

Figure 3.15 Chemical structure of β-carotene.

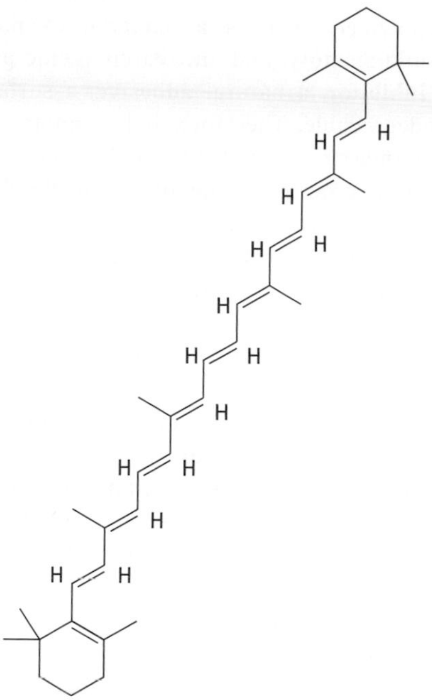

the early nineteenth century, Italy had the monopoly on citric acid production from lemons and limes. However, the First World War meant a decline in output and increase in price. Pfizer set up the first citric acid production plant in 1923 using the filamentous fungus *Aspergillus niger*. The compound (the structure is shown in Figure 3.16) is commercially produced in surface, submerged or solid state fermentation of *Aspergillus niger* grown at a pH 2.5–3.5.

Citric acid production can be improved by increasing the sugar concentration in the medium. A range of cheap carbon sources is suitable. Nitrogen limitation is necessary.

Figure 3.16 Chemical structure of citric acid.

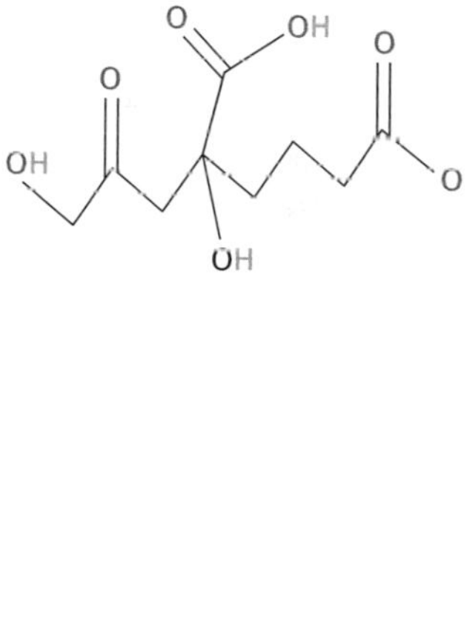

Key points

- The global market for industrial enzymes is worth billions of pounds.
- The biological detergent industry is the largest enzyme-using sector.
- Microbial exoenzymes (proteases, amylases, cellulases, lipases) for detergents are frequently produced in genetically modified microorganisms.
- Microorganisms are a major source of vitamins such as biotin, β-carotene, and vitamin D2, produced naturally in bacteria or fungi, as well as recombinant hosts.
- Citric acid is produced by the filamentous fungus *Aspergillus niger*.

3.3 Polymers

Microorganisms produce exopolymers for instance to facilitate aggregation or attachment. This can be to cells within a microbial population (such as in a biofilm), to solid surfaces or, in the case of pathogens, to host cells. Polymers inside microbial cells can function as energy reserves.

Polysaccharides

The fructose-homopolymer levan is a bacterial exopolysaccharide with anti-oxidant, anti-inflammatory, and anti-carcinogenic properties. It is also a **hyperglycaemic inhibitor**, a natural adhesive, a surfactant, and is biocompatible and biodegradable. Therefore, it has numerous applications in medicine, in the pharmaceutical, nutraceutical, cosmetic, food, and other industries, as well as a promising role in nanotechnology, e.g. for drug delivery. The enzyme levansucrase (sucrose-6-fructosyltransferase) catalyses the biosynthesis. Whole bacterial cells (e.g. *Erwinia amylovora, Pseudomonas syringae, Bacillus subtilis*) or isolated free and immobilized levansucrase are used for levan production. Different production strategies result in a variety of polymer attributes by design, and recombinant hosts will allow for large scale production.

Alginate has viscosifying and gelling properties, which make it suitable for use in the pharmaceutical industry (for instance, in hydrogels to promote wound healing), but also in the food, textile, and printing sectors. The polymer is comprised of variable ratios of β-D-mannuronate and its C-5 epimer α-L-guluronate linked by 1–4 glycosidic bonds. There are no branching or repeating blocks or unit patterns, unlike in xanthan. Originally it was isolated from brown algae *Laminaria* and *Macrocystis*, but microbial production processes based on bacterial synthesis pathways have since been developed. Mucoid *Pseudomonas aeruginosa* strains secrete alginate to support biofilms, and *Azotobacter vinelandii* produces a somewhat stiffer alginate when forming desiccation-resistant cysts.

› Another well-known microbially produced polysaccharide is xanthan gum, which is used as a thickening agent in salad dressings (amongst other foods). We will consider xanthan gum further in more detail in Chapter 4.

Bacterial cellulose is an unbranched polymer of β-1-4-linked glucopyranose residues; it is produced, for example, by *Gluconacetobacter xylinus* in batch fermentation using cheap carbon sources like date syrup. Cellulose is a thickener, stabilizer, and texture modifier. When used in the food industry it has the added benefit of being low calorie without flavour interactions.

Cyanophycin

Cyanophycin is a microbial storage polymer for nitrogen and carbon, and has a poly-aspartate backbone, where arginine is linked by the amino group to the carboxyl group of each aspartate, as illustrated in Figure 3.17.

Purified cyanophycin can be chemically polymerized. The polymer is a biodegradable substitute for synthetic polyacrylate in technical processes.

Figure 3.17 Chemical structure of cyanophycin.

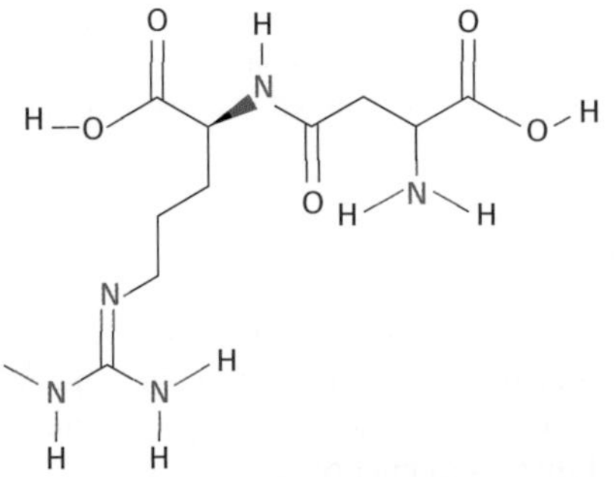

In physiological conditions, cyanophycin is located as insoluble granules in the cytoplasm, and for instance serves as a temporary nitrogen reserve in most cyanobacteria. However, cyanobacteria are not suitable for large-scale production. High productivity can be achieved in recombinant *Escherichia coli*, where up to one-quarter of the dry cell content is cyanophycin. Production takes place in the early exponential growth phase, with cyanophycin being extracted with acid from whole harvested cells.

Poly-γ-glutamate

Poly-γ-glutamate is a naturally occurring biodegradable, water-soluble, and edible biopolymer made of repeat units of L-glutamate, D-glutamate, or both. It has applications in the food, medical, and wastewater industries. *Bacillus* species and recombinant *Escherichia coli* are mostly used for production. Cheap substrates such as cane molasses, rapeseed meal, and corncob fibres can be used as a carbon source, but glucose is most suitable. Yeast extract is an excellent but costly nitrogen source, and $(NH_4)_2SO_4$ or NH_4Cl are alternatives. Secreted poly-γ-glutamate is recovered from the culture medium by precipitation or filtration.

Hyaluronic acid

Hyaluronic acid has the ability to retain large volumes of water. As such, it has applications as a medical biopolymer (e.g. corneal transplantation and retinal attachment surgery) and in cosmetics and speciality foods. The performance and therapeutic efficacy depend on the molecular weight of the polymer. As shown in Figure 3.18, the molecule is a linear glycosaminoglycan of varying length with glucuronate and N-acetylglucosamine joined alternately by β-1-3 and β-1-4 glycosidic bonds.

Hyaluronic acid is naturally synthesized as an extracellular capsule by pathogenic group A and C streptococci, and is commercially produced by group C streptococci (e.g. *Streptococcus equi*). Heterologous hosts (*Bacillus subtilis*, *Escherichia coli*, *Saccharomyces cerevisiae*) and metabolic engineering have been utilized, but usually recombinant systems have low productivity, and generally the product viscosity limits the yield. The productivity and molecular weight of hyaluronic acid are enhanced via intracellular ATP content.

Polyhydroxyalkanoates

Polyhydroxyalkanoates, including the well-known polyhydroxy butyrate, are linear polyoxoesters of hydroxyalkanoate monomers, as depicted in Figure 3.19.

Figure 3.18 Chemical structure of hyaluronic acid.

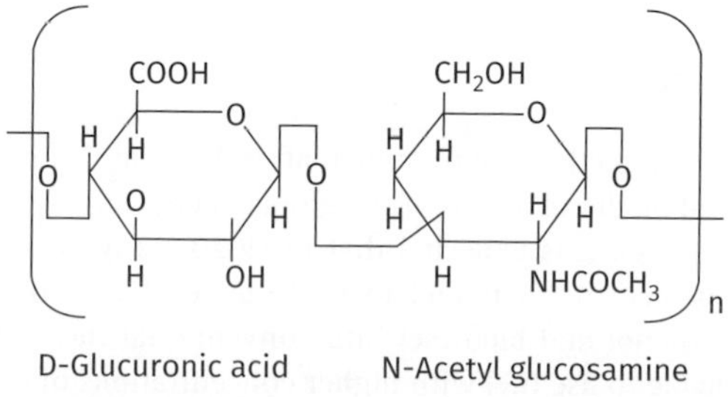

Figure 3.19 General structure of polyhydroxyalkanoates.

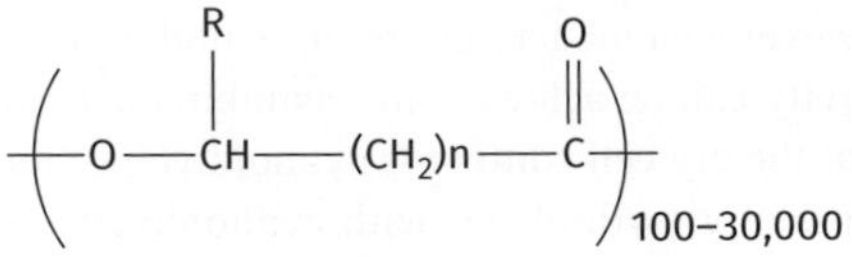

Given the ongoing depletion of fossil fuels, polyhydroxyalkanoates produced by microorganisms are suitable biological alternatives for plastic production, and moreover are biodegradable. Polyhydroxyalkanoate production is also linked to the biofuel industry. Bacteria store insoluble granules as a carbon source, and production can take place under phosphate or nitrogen limitation in a two-stage fed-batch process. Cheap substrates such as molasses, whey, wood, wheat bran, vegetable oils, waste cooking oils, glycerol, and even wastewater can be used. Initially a high bacterial cell density is achieved in nutrient-rich media, then polyhydroxyalkanoate production is increased by depleting a nutrient.

The range of natural producers is vast and includes hydrocarbon degraders (e.g. *Pseudomonas*, *Ralstonia*), halophiles (e.g. Archaea; *Halococcus*), photosynthetic bacteria (e.g. cyanobacteria *Synechococcus*), plant growth promoting rhizobia, and antibiotic producers (e.g. *Streptomyces*, *Bacillus*). Rather than using pure cultures and genetic or metabolic engineering, a recent innovative approach is to use mixed microbial consortia and evolutionary engineering. This means the environment rather than the microorganism is engineered by exerting selective pressure for a certain metabolism via feeding and cultivation conditions in the fermenter.

Key points

- The biocompatible and biodegradable fructose-homopolymer levan is produced by bacteria using the enzyme levansucrase.
- Bacteria produce the valuable polymers alginate, cellulose, and xanthan gum with high productivity.
- The microbial storage polymer cyanophycin can be obtained from cyanobacteria or at higher yield from recombinant hosts.
- The biodegradable, water-soluble, and edible biopolymer poly-γ-glutamate is produced in bacteria naturally or recombinantly.
- The medical biopolymer hyaluronic acid is commercially produced by group C streptococci and recombinant hosts.
- Polyhydroxyalkanoates are suitable biological alternatives for plastic production and can be obtained from bacteria grown on cheap substrates.

3.4 Biofuels

A biofuel is derived directly from organic matter. The main commercial biofuels are bioethanol and biodiesel. Much of the global energy supply still comes from fossil and nuclear fuel. It is expected that by 2020 many developed countries will require 10 percent of their fuel to be from renewable resources, mainly by blending bioethanol and biodiesel into conventional fuels. However, many older cars are unable to use fuel with higher concentrations of ethanol.

First-generation biofuels require large amounts of agricultural land to be used for the growth of corn, sugar beet, sugar cane, or other crops. As the land could be used instead to grow food for human consumption, such use is controversial. If the USA were to convert completely to corn-based biofuels, this would use up 70 percent of the current annual corn crop worldwide. As we cannot reduce the world food production nor increase the amount of agricultural land, feasible alternatives are required.

Second-generation biodiesel (see Figure 3.20) is generated from waste vegetable oils (unsuitable or previously used for cooking). Animal fat waste could be utilized for biodiesel. Advances in metabolic engineering and systems biology allow us to use cheap feedstocks such as sugar cane for the microbial production of compounds for advanced biofuels by organisms, including *Escherichia coli*, *Saccharomyces cerevisiae*, *Bacillus subtilis*, *Kluyveromyces marxianus*, and *Corynebacterium glutamicum*. Photosynthetic green algae are globally the fastest growing primary producers, and metabolic engineering allows altering the lipid composition and quantities, and to remove biofuel production bottlenecks related to harvesting, extraction, and conversion to fuel.

Microbial oil as a basis of biofuels

Microbial oil can be obtained from bacteria, fungi, and algae (e.g. *Aspergillus nidulans*, *Claviceps purpurea*, *Ustilago zeae*, *Yarrowia lipolytica*, *Rhodococcus opacus*, *Streptomyces coelicolor*, *Pseudomonas aeruginosa)*, with more than 20 percent of the biomass being lipids. Microbial oil extraction processes are simpler and cheaper than from plants. Depending on the fatty acid composition, the oils can be used in the biodiesel, pharmaceutical, cosmetic, food, and biopolymer industries. The composition can be controlled via the cultivation regime (e.g. aeration, pH, temperature, length), where the ratio of carbon and nitrogen source is crucial, usually with nitrogen limitation and excess carbon.

Figure 3.20 Approaches to biodiesel production. Starter materials and main steps are shown.

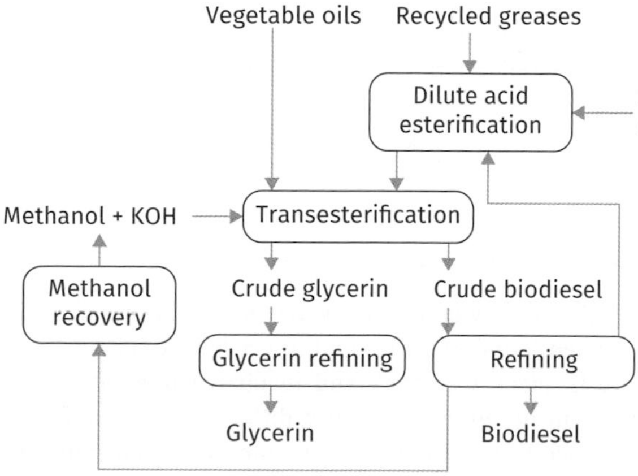

Adapted from Pinto, A. C., et al. Biodiesel: an overview. Journal of the Brazilian Chemical Society. J. Braz. Chem. Soc. vol.16 no.6b São Paulo Nov./Dec. 2005. Attribution-NonCommercial 4.0 International (CC BY-NC 4.0)

Ethanol as a basis for biofuels

Bioethanol can be generated from non-edible portions of plants (e.g. straws, brans, stalks, husks) that would otherwise be composted or turned into animal feed. The material needs milling, followed by the releasing of carbohydrates from lignocellulose and enzymatically digesting them into basic sugars that can be fermented into alcohol. This labour and energy-intensive process prevents bioethanol production at a competitive market price.

Microbial ethanol production is comparatively straightforward and uses a range of species. *Saccharomyces cerevisiae* and *Escherichia coli* for instance, ferment pyruvate from glycolysis under anaerobic conditions to ethanol. *Zymomonas mobilis*, however, uses the Entner-Doudoroff pathway instead of the Embden-Meyerhof-Parnas pathway to glycolysis and only produces one ATP molecule per glucose instead of two. More carbon is available for ethanol production, resulting in a 2.5-fold higher specific ethanol productivity than that of *S. cerevisiae*. Metabolic pathway engineering and physiological improvements of the producers will allow for more economical biofuel production.

Alkanes and alkenes as a basis of biofuels

A disadvantage of bioethanol is its low energy density compared with petrol-based fuels, whereas that of alkanes and alkenes is at par. Alkanes and alkenes are successfully produced in microorganisms following metabolic pathway engineering and the optimization of host and culture conditions. Initially, the metabolic pathway was identified in cyanobacteria. Recombinant hosts are *E. coli* and *S. cerevisiae*.

Propanol as a basis for biofuels

Isopropanol and n-propanol as biofuel components can be produced from renewable resources. Chemical conversion into propylene is required, which is still more expensive than from petrol.

Clostridia naturally produce propanol during anaerobic growth based on mixed acetone-butanol-ethanol fermentation.

The acetone-dependent isopropanol pathway in clostridia, which simultaneously produce isopropanol, butanol, and ethanol, is heterologously expressed in bacteria and yeast. In propionibacteria and *Clostridium propionicum* n-propanol is synthesized via the dicarboxylic pathway. *Clostridium acetobutylicum*, *Escherichia coli*, and yeast have been engineered to produce isopropanol. Advanced metabolic engineering allows for metabolic pathways to be expressed in various hosts.

Anaerobic digestion (biogas)

Anaerobic digestion is a widely used process to convert organic waste into biogas. Such waste is often found on farms and in wastewater plants. The sludge, slurry, and manure is added to a tank and incubated involving one or two stages. The process can be optimized to use different bacteria. Thermophilic anaerobic digesters operate at around 60°C and mesophilic anaerobic digesters operate at around 30–40°C. Manure and slurry are used in batch processes, while wastewater sludge is used in a continuous process. Under anaerobic conditions carbohydrates, proteins, and fats are gradually broken down by microorganisms to generate energy, as depicted in Figure 3.21. The lack of oxygen means that organic compounds cannot be fully broken down, and the end result of digestion is a mixture of methane and carbon dioxide. This gas can be siphoned off and used as biogas for vehicles, burnt in a generator to make electricity and heat, or burnt for heating.

The process requires a consortium of anaerobic bacteria. First, acidogenic bacteria carry out hydrolysis, acidogenesis, and acetogenesis, while methanogenic bacteria generate methane from the reduction of CO_2 or acetate.

For slurry a single phase digester (see Figure 3.22) is sufficient, however, for food and lignin-containing organic wastes acidification often lowers the rate of degradation and can inhibit methanogenic bacteria, limiting the rate of methane production. A two-phase digester is utilized for food wastes. In the first phase, acidogenic bacteria hydrolyse and acidify the mixture in a process that typically lasts 1–5 days. The hydrolysed material is then fed into another digester for the second, methanogenic, phase that can last up to 30 days. The separation of the two phases allows for better control of pH in the methanogenic phase, allowing for different food wastes to be used.

Figure 3.21 The steps of anaerobic digestion. Depending on the initial biological molecule, hydrolysis leads to different compounds.

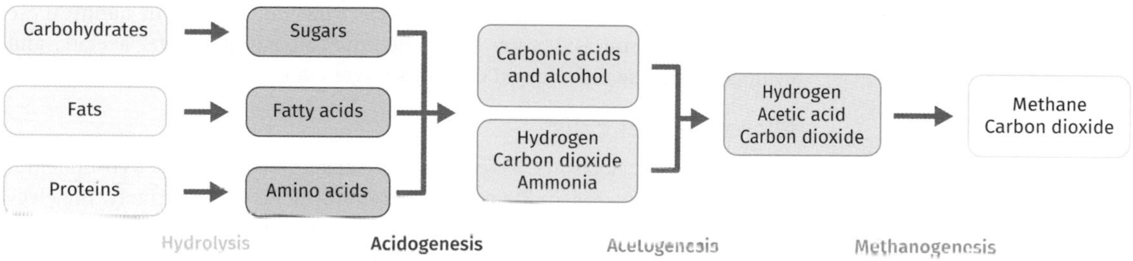

Figure 3.22 Single-and two-phase anaerobic digestion. Separation allows the use of different food wastes.

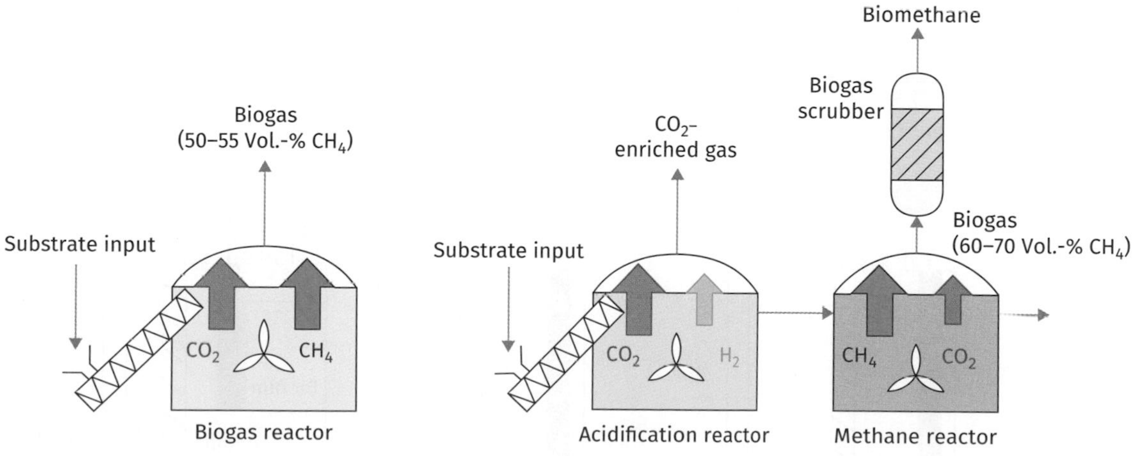

Queens University Belfast.

Key points

- Bioethanol and biodiesel are the main commercial biofuels.
- Metabolic engineering allows using cheap substrates for microbial production of compounds for advanced biofuels.
- Microbial oil as a basis of biofuels can be obtained from bacteria, fungi, and algae at high yield with a composition controlled by the cultivation regime, and then followed by simple extraction processes.
- Microbial ethanol production uses a range of species and optimized processes result in increased productivity.
- Alkanes and alkenes produced in microorganisms have an energy density comparable with petrol-based fuels.
- Propanol as a basis of biofuels is produced in various optimized hosts.
- Anaerobic digestion converts organic waste into biogas by a consortium of acidogenic and methanogenic bacteria.

3.5 Bioelectrogenesis

A large number of bacterial species produce electricity as they break down substrates. These bacteria are known as electrogenic bacteria. In nature these bacteria live in soil and water sediments, oxidize, e.g. pyruvate, acetate, and hydrogen, and use the released electrons to generate a proton motive force.

Microbial fuel cells

Bacteria release electrons which can be transported to solid environmental substrates such as iron-oxide minerals or humic acids. Alternatively, the electrons can be transferred to an electrode. A current will flow and electrical power will be generated if the electrode is part of a circuit. This construct is called a microbial fuel cell. There are two types of basic microbial fuel cells: single-chamber and dual-chamber fuel cells (see Figure 3.23). Both types use two electrodes: an anode that collects the electrons from the bacteria and a cathode where electrons are released by reducing an electron acceptor

Figure 3.23 Single-chamber and dual-chamber microbial fuel cells. Their performance varies.

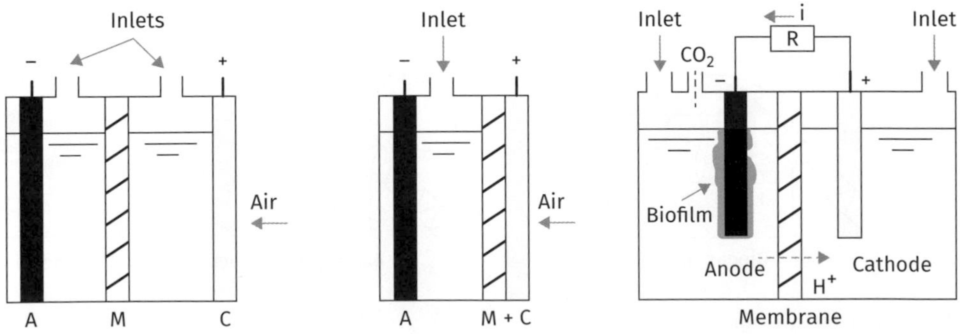

(typically turning oxygen into water). The electrons move from the anode to the cathode driven by the redox potential difference (voltage) between the reducing bacteria on the anode and the chemical reaction on the cathode. A membrane allows the transfer of positive charge between the electrodes, completing the circuit.

Physiology of electrogenic bacteria

Geobacter sulfurreducens and *Shewanella oneidensis* MR-1 are the most extensively studied model electrogenic bacteria.

Bacteria transfer electrons into the anode by two mechanisms:

- the direct method
  - the cell surface touches the anode (Figure 3.24(a)) or
  - conductive pili connect to the anode like molecular wires (Figure 3.24(b));
- the indirect/mediated method
  - cells reduce electron shuttles that diffuse between the bacterium and the anode (Figure 3.24(c)). These electron shuttles can either
    - be secreted by the cell (endogenous mediators) or
    - added to the culture (exogenous mediators).

Figure 3.24 Electron transfer mechanisms from bacterial cell to anode.
The transfer is either direct (a, b) or indirect (c).

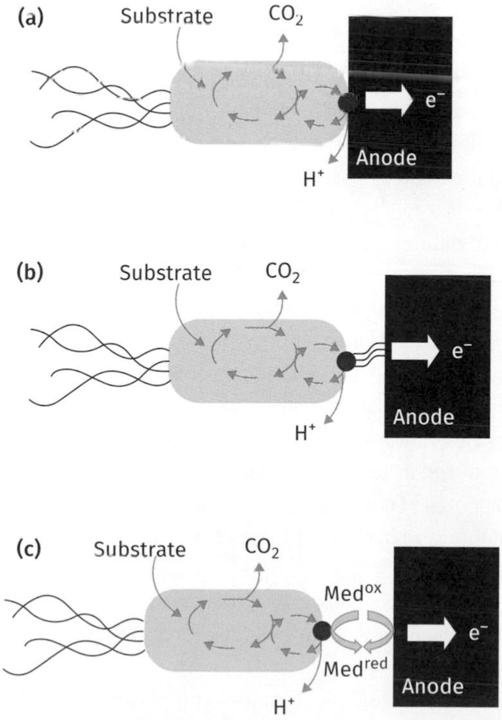

Biofilms that coat the anode tend to direct electron transfer, whereas free-swimming (planktonic) bacteria typically use mediated electron transfer. The most productive microbial fuel cells use microbial consortia for long-term efficient electricity production. Despite many advances, the power output of microbial fuel cells is only now beginning to compete with alternative power sources such as solar or wind. Weather buoys, electricity-generating portable urinals, and pilot-scale treatment of brewery wastewater are examples of applications of microbial fuel cells.

Lowering the potential on the cathode side of the microbial fuel cell drives the electrons back into the bacterial cell, where they can be used in reductive anabolic reactions. This process is called bioelectrosynthesis and has been successfully conducted at laboratory scale for instance as CO_2 fixation to formate and acetate.

Key points

- A large number of bacterial species produce electricity based on establishing a proton motive force.
- The released electrons can be transported to solid environmental substrates or an electrode, creating a current if part of a circuit.
- The flow of electrons can be reversed in a process called bioelectrosynthesis.

Chapter Summary

- Microorganisms produce compounds worth many billions of pounds each year, directly or by transforming chemical compounds.
- Screening for antibiotic producers can be based on isolated microorganisms or culture-independent methods using isolated DNA.
- Microbial biotransformation of natural steroids is used to produce steroid drugs.
- Production of therapeutic proteins has seen considerable advancement since the initial creation of recombinant vectors.
- The global market for industrial enzymes is worth billions of pounds. The biological detergent industry is the largest enzyme-using sector.
- Microbial exoenzymes (proteases, amylases, cellulases, lipases) for detergents are frequently produced in genetically modified microorganisms.
- Microorganisms are a major source of vitamins such as biotin, β-carotene, and vitamin D2, produced naturally in bacteria or fungi as well as recombinant hosts.
- Citric acid is produced by the filamentous fungus *Aspergillus niger*.
- Bacteria produce the valuable polymers alginate, cellulose, and xanthan gum with high productivity.
- Polyhydroxyalkanoates are suitable biological alternatives for plastic production and can be obtained from bacteria grown on cheap substrates.

- Metabolic engineering allows the use of cheap substrates for microbial production of compounds for advanced biofuels.
- Microbial oil as a basis of biofuels can be obtained from bacteria, fungi, and algae at high yield with a composition controlled by the cultivation regime, and then followed by simple extraction processes.
- Anaerobic digestion converts organic waste into biogas by a consortium of acidogenic and methanogenic bacteria.
- A large number of bacterial species produce electricity based on establishing a proton motive force.

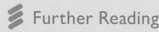

Further Reading

Cheon, S., et al. (2016). 'Recent trends in metabolic engineering of microorganisms for the production of advanced biofuels'. Current Opinion in Chemical Biology 35:10–21.
The following review gives a good overview of approaches to advanced biofuel production.

Sanchez-Garcia, L. et al. (2016). 'Recombinant pharmaceuticals from microbial cells: a 2015 update'. Microb Cell Fact 15:33, DOI 10.1186/s12934-016-0437-3.
The following paper discusses updates to recombinant pharmaceutical production.

Singh, R. et al. (2016). 'Microbial enzymes: industrial progress in 21st century'. Biotech 6:174, DOI 10.1007/s13205-016-0485-8
The following paper reviews microbial enzyme production.

Tharali, A. D., Sain, N., and Jabez Osborne, W. (2016). 'Microbial fuel cells in bioelectricity production'. Frontiers in Life Science 9:4 252–66 http://dx.doi.org/10.1080/21553769.2016.1230787
The following paper gives some insights into microbial fuel cells.

zu Berstenhorst, S. M., Hohmann H-P., and Stahmann, K-P. (2009). 'Vitamins and vitamin-like compounds: microbial production'. Encyclopedia of Microbiology 549–61, DOI 10.1016/B978-012373944-5.00161-9.
The following paper reviews microbial vitamin production.

Discussion Questions

3.1 Discuss the difference bio-production makes to our lives.

3.2 Discuss the role of genetic engineering in the context of bio-production.

3.3 Discuss the role of microorganisms in the context of global pollution with plastic.

3.4 Discuss the future of biofuels.

4

BIOTECHNOLOGY AND FOOD AND DRINK PRODUCTION

Learning Objectives

- To put in context that humans have had a long and rich history of their use of microbes for food production, initially for preservation;
- to explain how microbes can be used to make components for food production, including amino acids, enzymes, oils, and polysaccharides;
- to describe how microbes are used to make food directly, examples include tempeh, fermented meat products, cheese, and bread;
- to explain how microbes are used to make beverages both alcoholic, such as beer and wine, and non-alcoholic, such as coffee and cocoa;
- to explain how microbes can be used as food in a variety of ways, including single-cell protein;
- to explain the role of microorganisms in the production of food supplements such as vitamins, pigments, artificial flavours, and probiotics;
- to be able to give an overview of the role microorganisms play in the context of food and beverages;
- to be able to describe with examples that many principles apply to both food and beverage production;
- to be able to discuss the use of microorganisms in context of past, present, and future nutritional challenges;
- to be able to identify some of the microbial input in foods and beverages we come across.

It would be hard to go through your day without coming across the influence of microbes in the foods and beverages you eat and drink. Humans have had a long history of using microbes both to produce and to preserve food; in fact, preservation and mummification were related technologies. The changes we see in foods when they are naturally or deliberately preserved are a result of changes in pH, and the production of gas and ethanol which limits microbial growth, thus allows preservation and increases the gastronomic value of the food. Preservation can also lead to changes in the texture of food, which can add to their 'shelf life', for example, the precipitation of proteins when milk is turned into cheese. Cheese was first thought to have been made 8,000 years ago in the Middle East. Ancient Sumerians also made a bread-beer. The production of wines resulted from the close proximity of the yeast living on the grape; thus, fermentation occurred without the addition of any further yeast.

Despite the use of these food and beverage fermentation methods, ancient people did not really know that such methods used microbes. It wasn't until the 19th century that scientists such as Louis Pasteur [1822–1895] and Robert Koch [1843–1910] began to systematically show the role played by microbes in processes such as brewing (in addition, of course, to their role in disease). This chapter will explore how microbes are used in the food production industry: some uses are obvious, others much less so.

4.1 Microbes producing components for food

Microbes are often used to make components which are then added to foods. In this section we will consider some of these key food components: enzymes, lipids, and polysaccharides, but we will begin with one of the biggest industries—the production of amino acids.

Amino acids

Amino acids are required as the building blocks of proteins; some are known as essential amino acids as they cannot be made by the body, and must be supplied by the diet. In cells, amino acids are made from intermediates of the citric acid cycle and other major metabolic pathways. For a deeper understanding of this area we recommend the reading of the Biochemistry primer in this series. Amino acids are produced commercially from microbes on a huge scale, with five million tonnes of various amino acids being made every year. Amino acids are used for a wide range of products. For example the amino acid, aspartic acid, is used to make aspartame, a low-calorie sweetener. However, the production of L-glutamic acid (GA) is the largest industry, with over a million tonnes produced annually. L-glutamic acid is mainly used as a seasoning in the form of monosodium glutamate (MSG), but it is also used as a starting point to make other chemicals. Originally, it was produced by the hydrolysis of wheat gluten or soybean protein, but now GA is produced by fermentation.

The reaction scheme summarizing the production of GA is shown in Figure 4.1; notice that it requires the action of the enzyme glutamate dehydrogenase. The organisms used in fermentation to produce GA are Gram positive, non-spore forming, and non-motile bacteria; examples are given in Table 4.1.

Table 4.1 Bacterial species used for L-glutamic acid production

Genus	Species
Corynebacterium	*C. glutamicum*
	C. herculis
Brevibacterium	*B. flavum*
	B. lactofermentum
	B. roseum
Microbacterium	*M. flavum*
Arthrobacter	*A. globiformis*

Corynebacterium glutamicum has been used extensively over many years. First isolated in Japan, it grows easily, can be readily manipulated through genetic techniques, and is generally regarded as safe (GRAS; see Chapter 1). The carbon source for the fermentation is molasses or sugarcane juice; more complex organic wastes are avoided, as their presence makes it hard to purify the glutamic acid post-fermentation. Other nutritional supplements are added to the fermentation; these include yeast extract and peptone.

The fermentation is an aerobic batch process in stirred reactors with volumes of around 450 m^3. The temperature used depends on the growth properties of the producing organism but is usually between 30–37°C. Fermentation lasts for 35–40 hours, and glutamic acid starts to accumulate half way through this time. The nitrogen source is ammonium or urea, but this has to be carefully controlled as too much will reduce the level of GA produced.

The recovery of GA is quite a complex process. The fermentation culture is filtered and the calcium glutamate is bleached using activated charcoal. The calcium glutamate is then evaporated to a concentration of 37 percent at 70°C and 0.57 atm. Acidification then occurs with sulphuric acid, the resulting calcium sulphate precipitate is removed, and glutamic acid is then further evaporated and then crystallized. Monosodium glutamate is prepared by the addition of sodium hydroxide to GA crystals. The MSG is then re-crystallized.

While glutamic acid represents the biggest production of an amino acid, L-lysine is the fastest growing market. L-lysine is used as an animal feed additive, mainly for poultry, but increasingly as a fish feed supplement. L-lysine is produced on a large scale using the bacterium *C. glutamicum;* 50 percent of the market is in the USA and China, where maize is used for the fermentation.

Enzymes

Of the enzymes produced on an industrial scale, over half of them come from fungi, and one-third come from bacteria. There are six classes of enzyme that are related to food, which have different modes of reaction; these are detailed in Table 4.2. Enzymes are important components of many different

Figure 4.1 Equation for the production of glutamic acid.

$$\text{Oxoglutaric acid} + NH^{+}_{4} + NAD(P)H + H^{+} \rightarrow \text{L-glutamic acid} + DAD(P)^{+} + H_2O$$

products, a good example being those found in washing powders. These include proteases produced by *Bacillus* spp. (e.g. *Bacillus licheniformis*), which break down proteins into polypeptides and amino acids (see Chapter 3). Enzymes are used extensively in the food industry. Look at Table 4.3 to see some more examples of enzymes, their producing organisms, and their uses in the food industry. Notice how their uses are very wide-ranging; some may even surprise you!

Improving enzyme function

Genetic engineering can be used to modify the structure of an enzyme, using either random or targeted mutation, to improve or alter its activity (see Chapter 1). We call such a process protein engineering. Targeted mutation (often called 'rational

Table 4.2 Enzyme classes as related to food, mode of reaction, and examples

Class	Mode of reaction	Industrial enzyme
Oxidoreducatase	Oxidation reactions involving the transfer of electrons from one molecule to another	Catalase Glucose oxidase Laccases
Transferases	Catalyse the transfer of atoms from one molecule to another	Fructosyltransferases Glucosyltransferases
Hydrolases	Catalyse the cleavage of substrates by water	Amylases Cellulases Lipases Mannases Pectinases Phytases Proteases Pullulanases Xylanases
Lyases	Catalyse the addition of groups to double bonds or the formation of double bonds through the removal of groups	Pectate lyases Alpha-acetolactate Decarboxylases
Isomerases	Catalyse the transfer of groups from one position to another on the same molecule	Glucose isomerases Epimerases Mutases Lyases Topoisomerases
Ligases	Catalyse the joining of molecules together with covalent bonds	Argininosuccinate Glutathione synthase

Table 4.3 Microbial-produced enzymes and their uses in the food industry

Enzyme	Example of producing organism(s)	Use
Amylase	*Bacillus licheniformis* *Aspergillus oryzae*	Hydrolysis of starch in the bread and beer industry
Cellulase	*Trichoderma reesei* *Bacillus* spp. *Clostridium* spp.	Production of fruit drinks
Invertase	Yeasts, *Aspergillus* spp.	Hydrolysis of sucrose. Used to make soft centres in chocolates
Lactase	*Aspergillus niger* *Aspergillus oryzae*	Hydrolysis of lactose in milk products
Lipases	*Bacillus* spp. *Candida rugosa*	Used to modify food flavour by the synthesis of esters Refining rice flour and modification of soybean milk
Pectic enzymes	*A. niger* *P. notatum*	Pre-treatment of fresh juices
Protease	*Aspergillus* spp. *Bacillus* spp.	Tenderize meat, rennin replacement
Phytase	*Lactobacillus casei* *Bacillus badious*	Producing animal feed stocks
Rennet	*Cryphonetrica parasitica*	Milk coagulation in cheese manufacture

design') uses site-directed mutagenesis to create specific amino acid substitutions. The use of molecular genetics has enabled the cost of enzyme production to be reduced, and enzymes can be made in host organisms which are safe to cultivate on a large scale.

Microbial oils

Many microbes can produce a large amount of their dry weight as lipid (or oil) under known cultivation conditions. Those microbes with a lipid content greater than 20 percent biomass weight are known as oleaginous. Oil accumulation by oleaginous microbes requires nutrient-limited conditions; nitrogen is not added to the fermentation media so that excess carbon is used to make storage oils and not new cells. Due to their fast growth, high-cell density, and good oil production, microalgae are often targets for the development of microbial oils. The most commonly used is *Chlorella*. The company TerraVia use algae grown on sugarcane, corn, and stover to produce oils for food, as well as animal nutrition. Microbial oil produced for the food industry has to be extracted under stricter conditions than those used to produce biofuel (remember what you learnt in Chapter 3).

Microbial polysaccharides

Microbial polysaccharides can come from a range of different microbes including bacteria, yeasts, moulds, and algae. In Figure 4.2 you can see the shiny yellow colonies of *Xanthomonas campestris* which produces xanthan gum. Many (but not all) microbial polysaccharides are water-loving and disperse easily in water. They can be made up of a single monosaccharide (in which case they are called homopolysaccharides), or can comprise of several different monosaccharides (and so are called heteropolysaccharides). Microbial polysaccharides are classified into three classes, depending on where they are located in the cell:

1. Intracellular polysaccharides are located on part of the cytoplasmic membrane
2. Cell wall polysaccharides form a structural part of the cell wall itself
3. Extracellular polysaccharides are loosely attached to the cell surface in the form of slime or covalently bound to the cell surface to form a capsule.

Microbial polysaccharides have an important natural role in the development of biofilms, where they help cells to adhere to surfaces and enhance resistance to environmental conditions.

Microbial polysaccharides can be used in a number of different industries—food, pharmaceutical, and also engineering—some of which we already discussed in Chapter 3. Table 4.4 shows bacteria which produce commercial polysaccharides, their classification, and how they are used. However, most microbial polysaccharides have to be slightly modified, either chemically or physically, before they can be used. This is because their properties often do not meet the needs of polysaccharides used in the food industry.

Figure 4.2 Colonies of *Xanthomonas campestris*.

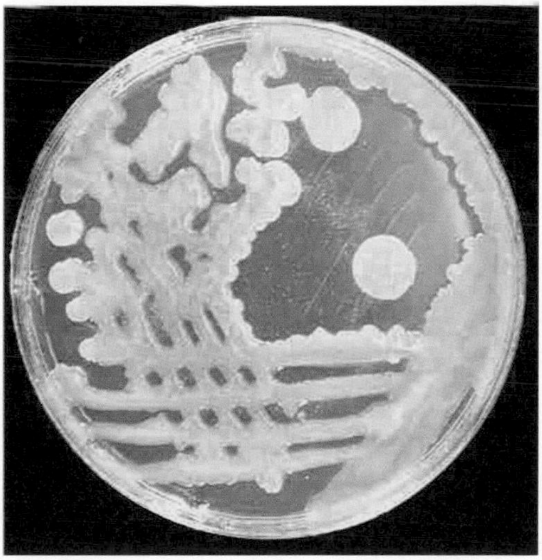

Table 4.4 Commercial microbial polysaccharides, their classification, and how they are used

Bacteria	Polysaccharide	Class	Group	Use
Agrobacterium spp.	Curdlan	Exopolysaccharide	Homopolysaccharide	Foods, pharmaceutical industry, heavy metal removal, concrete additive
Halomonas eurihalina	Levan	Exopolysaccharide	Homopolysaccharide	Prebiotic, pharmaceutical and cosmetic industries
Leuconostoc dextranicum	Glucan	Cell wall	Homopolysaccharide	Pharmaceutical and food industries
Leuconostoc mesenteroides	Dextran	Exopolysaccharide	Homopolysaccharide	Pharmaceutical industry and thickener in food
Gluconacetobacter xylinum	Cellulose	Cell wall	Homopolysaccharide	Pharmaceutical in wound healing, tissue engineering, and audio speaker diaphragms
Pseudomonas putida	Alginate	Cell wall	Heteropolysaccharide	Food industry, textiles, and paper making
Sphingomonas paucimobilis	Gellan	Exopolysaccharide	Heteropolysaccharide	Confectionary and jams
Xanthomonas spp.	Xanthan gum	Exopolysaccharide	Heteropolysaccharide	Food industry as a stabilizer and thickener. Ceramic, cosmetic and textile industry. Also used as a lubricant for oil drills

Key points

- Microbes produce components which are used in the manufacture of other foods and drinks.
- The production of amino acids, such as glutamic acid is done on a large scale.
- There are a number of different enzymes which are produced by microbes which are used to make a range of food and drink products. Enzyme function can be improved by the application of deliberate mutation.
- Microbial oils intended for use in food have to be extracted under strict, controlled conditions
- Microbial polysaccharides can be classed into three different groups. Usually they have to be chemically or physically modified before they can be used in the food industry.

4.2 Microbes used to make food

Who does not love food? It might surprise you, but without microorganisms we would have a much shorter and relatively boring menu to choose from. We mentioned in the introduction that our use of microorganisms to make food was initially more of a coincidence than deliberate. Microbial activity frequently prolongs shelf life, but it also changes food consistency and flavour. Food also becomes more digestible and enriched with vitamins and minerals. Much of this input is based on fermentations which is the obtaining of energy from anaerobic oxidation of reduced organic substances (chemo-organotrophy). Such substrate-level phosphorylation produces energy (ATP).

During fermentations, microorganisms transform the chemical components of the food directly as substrates, or indirectly as a consequence of the changed chemical nature (e.g. fats may become oxidized, a lowered pH results in protein precipitation). The mixed populations of prokaryotic and eukaryotic microorganisms involved are often part of a food web, where the fermentation product or by-product of one species is the substrate of the next. The cells also produce other compounds as part of their entire metabolism, adding food-typical flavours. That is why the flavour of a given food depends on specific incubation conditions such as temperature and time. It is important to get these right, particularly for high-quality or even protected-origin foods.

Vegetables and other plants

Fermented vegetables, like cabbage, are a popular and traditional food around the world—from sauerkraut in Eastern Europe to kimchi in Korea. The pickle-like taste is partly due to the production of lactic acid by Gram-positive lactic acid bacteria such as *Leuconostoc* sp. and *Lactobacillus* sp. *Lactobacillus acidophilus* produces the lactic acid homofermentatively as you can see in Figure 4.3. 'Homofermentative' tells us that lactic acid is the only product of the fermentation. We will consider the heterofermentative pathway, yielding more products than lactic acid, later in the context of dairy products.

This basic principle of fermenting food using lactic acid bacteria also applies to the production of silage as animal food, during which grass and other plant material is fermented. Fermentation processes can produce complex flavours in food, which cater for acquired tastes. Olives would be such an example; their production process is varied and can include fermentation in brine. *Lactobacillus pentosus* and *Lactobacillus plantarum* are frequently in the starter culture for green olives when produced using the most common method (Spanish style). Yeasts are also often present in the culture, but less prevalent than in Greek style olives.

Tempeh or tempe (see Figure 4.4) from Indonesia is an example of using fungi to make food. The mycelium of *Rhizopus oligosporus* is grown in and around soybeans which have been previously cooked, and the fungal enzymes hydrolyse the soybean constituents. As a result, everything comes together as a compact cake due to the binding effect of the fungal hyphae. This soybean-based food should not be confused with tofu, which rather resembles the production of cheese, as we will discuss later, and starts with hot soy milk that gets coagulated.

It is also worth mentioning the production of vinegar here. Vinegar is not a food, but a condiment used to preserve and flavour foods. We could also discuss it in the beverages section later, because vinegar was actually drunk

Figure 4.3 The fermentation of glucose in homofermentative and heterofermentative lactic acid bacteria.

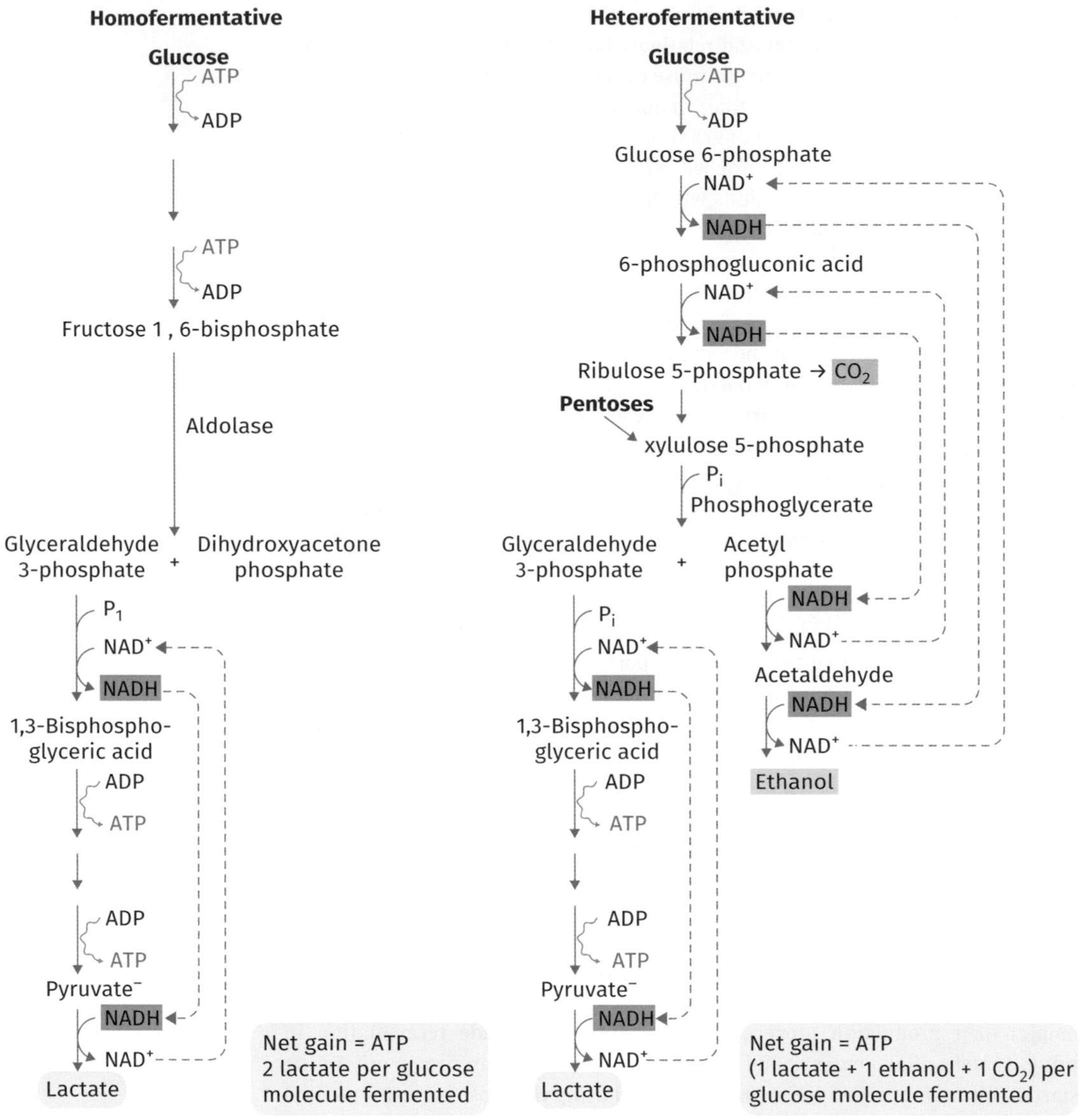

back in the early days of its production more than 4,000 years ago. In Egypt and Mesopotamia, wine and beer were left in open vessels. This exposure to air caused the ethanol to be oxidized to acetic acid (see Figure 4.5), resulting in the formation of vinegar.

Flavours of vinegar depend on the starting material such as red wine for red wine vinegar and cider for cider vinegar. Some *Acetobacteraceae* oxidize alcohols and sugars to organic acids. Oxygen is required for oxidation, and thus some microbes, for instance *Acetobacter xylinus*, form a biofilm of bacterial cellulose to float on the surface of the liquid. This principle was applied in the 'open-vat method' (Orleans method) of larger-scale commercial

Figure 4.4 Tempeh.

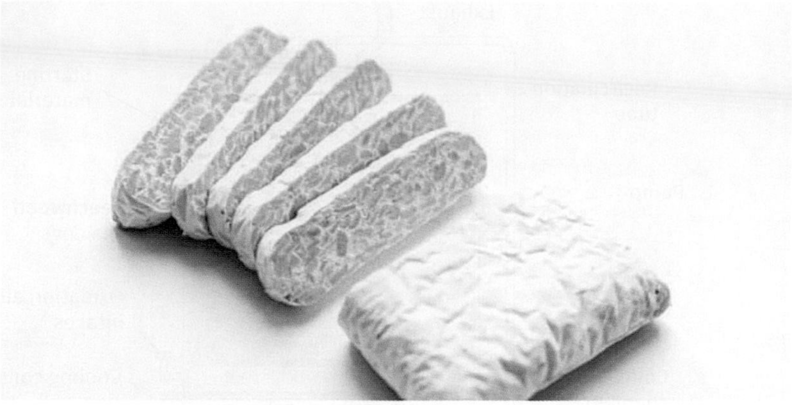

dani daniar/Shutterstock.com

Figure 4.5 Vinegar production *UQ*. The enzyme alcohol dehydrogenase oxidizes ethanol to acetaldehyde and reduces ubiquinone to ubiquinol UQH_2, and subsequently acetaldehyde to acetic acid by aldehyde dehydrogenase. The energy generated by the transfer of protons further in the chain is used for the production of ATP.

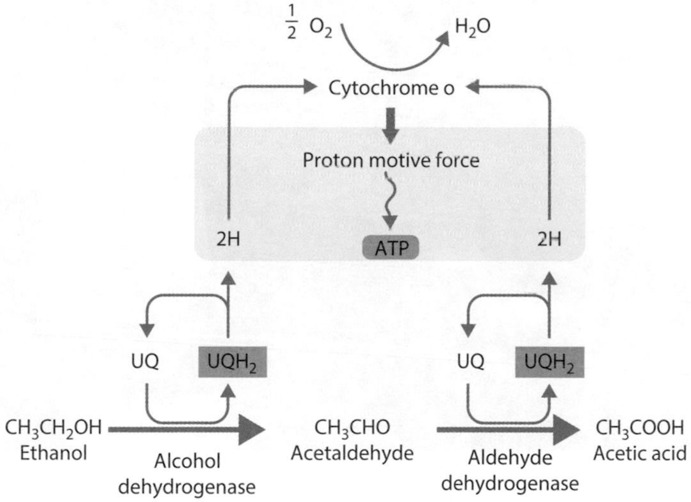

vinegar and acetic acid production. Wine was placed in large but shallow vats with the biofilm on the surface. The more efficient 'trickle method' (quick vinegar method) uses bacteria in a protective slime layer on wood shavings in a vinegar generator. As shown in Figure 4.6, the wine trickles over the inoculated shavings. The more efficient bubble method uses submerged fermentation (as discussed in Chapter 1.)

The vinegar produced can be further distilled to make it a more concentrated acetic acid. This concentrated acid does not add flavour to food and is usually just used to lower its pH like in pickling. Low-quality ethanol can be used as

Figure 4.6 **A vinegar generator.**

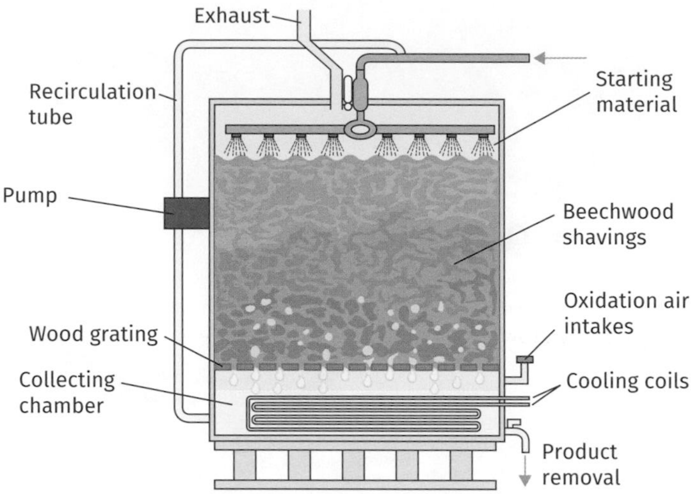

Figure 4.7 **Aceto balsamico wooden barrels.**

giovannifederzoni/Shutterstock.com

initial substrate in this production process. On the other end of the spectrum we have the famous example of the origin-protected types of aceto balsamico. These balsamic vinegars are aged and matured for many years in wooden barrels, as seen in Figure 4.7.

Soy sauce is another fermented food condiment you are probably familiar with. The fungus *Aspergillus oryzae* ferments a mixture of soybeans and some wheat (the amount depends on the region). This step is followed by a brine fermentation at high salt concentrations using halophilic bacteria and osmophilic yeasts.

Meat products

The basic principles of microbial activity when producing food from plants also apply to meat (usually pork and beef), and again some foods are origin-protected such as Parma Ham. For instance, salami sausage is produced by adding a starter culture (lactic acid bacteria and coagulase-negative cocci) to the meat and seasoning. Over time, the pH drops to less than 5, leading to a change in the structure and taste of the meat.

Often these meat products are then smoked and dried. Depending on geographical area, starter cultures and production processes vary. For example, European fermentation is slower (*Micrococcus* sp.) compared with the rapid US process (*Lactobacillus plantarum*, *Pediococus acidilactici*). Mould may also be purposefully grown on the product, as is the case with the growth of *Penicillium nalgiovense* on some Italian salamis.

The shelf life of cured meats is extended due to the combination of nitrite, salt, and the lowered pH (due to lactic acid bacteria). The nitrite not only reduces the oxidation of lipids, but also inhibits growth of the Gram-positive anaerobic spore-forming bacterium *Clostridium botulinum*, which is often brought in with spices and can cause food poisoning. The infection can even lead to death, because *Clostridium botulinum* produces the neurotoxic botulinum toxin. Monitoring for *Clostridium botulinum* is especially necessary when conditions are near anaerobic. This can be the case when fish products are fermented following longstanding traditions in Southeast Asia and Northern Europe, where bacterial fermentation is combined with the activity of natural enzymes present in the fish intestinal tracts.

Dairy products

The production of cheese in the Middle East dates back anything up to 8,000 years. In the early production of cheese, curdling was induced by directly using stomach content from young animals. Suckling animals curdle ingested milk enzymatically as part of their normal diet. The protease rennin (also known as chymosin) in the stomach of calves clots milk by cleaving the Ser-Phe^{105}+ Met-Ala bond in the kappa-chain of the milk protein casein.

A crucial substrate for microorganisms in milk is the disaccharide lactose, composed of galactose and glucose. We saw in Figure 4.3 that the fermentation of glucose results in the production of lactic acid. Over time the pH in milk drops from neutral to less than 5.5 in cheese, and even lower in other fermented milk products. This low pH curdles milk and some cheeses (called sour milk cheeses) like cottage cheese are produced without rennin. However, this is not the major approach when setting out to separate the solid (curd) and liquid (whey) part of milk.

There are more than 2,000 types of cheeses from across the world. These cheeses vary according to characteristics such as:

- the animal from which the milk was sourced;
- whether or not the milk has been pasteurized;
- its fat content—from cream to full fat, to semi-skimmed (parmesan) to whey (ricotta);
- the production process and temperature;
- the length of ageing before consumption;

The bigger picture panel 4.1

Please let us know of any dietary or ethical requirements

Commercially available rennet is a mixture of several enzymes related to milk digestion which is extracted from the stomachs of butchered calves. However, this source is not sustainable for large-scale manufacturing and also has ethical implications. In addition, religious beliefs may preclude using bovine products. As an alternative, bovine rennin has been produced using recombinant DNA technology in *Escherichia coli*, in *Bacillus* sp., yeasts, and filamentous fungi (recall the explanations around recombinant protein production in Chapters 1 and 3). While this approach does not use proteins extracted from animals, it technically is animal protein and would not necessarily be acceptable to vegetarians. Likewise not everyone wishes to ingest anything originating from genetically modified organisms. Fortunately, a variety of microorganisms such as the fungi *Mucor* sp. and *Endothia* sp. naturally produce milk-coagulating enzymes which can be used in cheese production. This example illustrates the multitude of challenges related to food production and the need for full and correct labelling.

What are the issues in this complex situation, and are there any that have not yet been alluded to? What are the challenges in addressing any of the issues?

- whether the rind is mould-ripened (Camembert: *Penicillium camemberti*) or washed for bacterial activity on the rind (*Brevibacterium linens*);
- whether it is veined with mould (Roquefort: *Penicillium roqueforti*).

The microorganisms found in the milk or on the skin of the animals or the dairy farmers were the original starter cultures. Naturally, lactic acid bacteria would make up about 80 percent of all milk flora, but the very clean nature of modern milk production processes means that the starting bacterial load is very low and starter cultures are needed. There are single-strain starters (e.g. *Streptococcus cremoris*), multi-strain starters (three or more single strains of *S. cremoris* and/or *S. lactis*), and mixed-strain starters (mixtures of strains of *S. cremoris*, *S. lactis*, *S. diacetylactis*, and *Leuconostoc* sp.). Given this variety, the often highly diverse cheese flavours (and odours!) should not come as a surprise. These flavours are developed during the ripening process and are determined by the particular acids, esters, diacetyl, and sulphur-containing compounds found in the cheese.

The microbial food web in dairy products adds further flavours and features. *Propionibacterium freudenreichii* secondarily ferments lactic acid to propionic acid, acetic acid, and carbon dioxide. The propionic acid contributes to flavour, whereas the carbon dioxide becomes trapped during maturation, generating holes in the cheese (e.g. Emmental). Cultured butter production starts with pasteurized cream and bacteria such as *Lactococcus lactis* and *Leuconostoc citrovorum* are added. *Leuconostoc mesenteroides* ssp., *Cremoris*, a heterofermentative lactic acid bacterium, produces diacetyl, which gives butter its distinctive flavour. Heterofermentative bacteria produce ethanol and carbon dioxide in addition to lactic acid.

Other heterofermenters such as *Lactobacillus delbrueckii*, *Lactobacillus acidophilus*, and *Streptococcus thermophilus* are used to produce yoghurt. Making yoghurt at home (Figure 4.8) is not that different to large-scale production in the dairy industry (Figure 4.9). Milk is heated and cooled. Next the

Figure 4.8 Making yoghurt in the kitchen – The most basic approach.

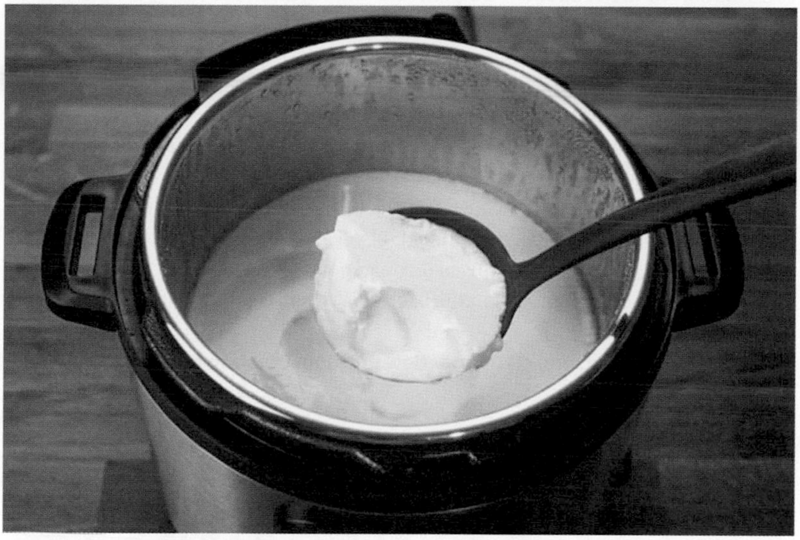

Tanya Stolyarevskaya/Shutterstock.com

Figure 4.9 Yoghurt-making machine in industry.

branislavpudar/Shutterstock.com

starter culture is added. At home this is initially a small amount of commercial yoghurt with living cultures, but subsequently we can use some of our own produce. The yoghurt is then incubated at 40–45°C for several hours or overnight, then cooled ready to enjoy.

The dairy industry, apart from producing delicious foods, makes an invaluable contribution to the economy and provides employment for many individuals in addition to microbiologists. The same applies to bread and beverage production, which we will cover now.

Bread

Leavened bread has been made by human beings for thousands of years. Figure 4.10 and Figure 4.11 show related artefacts.

The open texture of bread is caused by the release of carbon dioxide by the yeast *Saccharomyces cerevisiae*. This organism, also called baker's yeast, belongs to the same species as brewer's yeast (used for alcoholic fermentation), but is a different strain. Flour contains yeasts and bacteria. When we add water the microbial

Figure 4.10 **Bread, fruit, and grain from New Kingdom Egyptian tombs, ca 1400 BC.**

Figure 4.11 **Bakers making dough and filling bread moulds.** Painting; tomb of Qenamun, West Thebes, Sheikh Abd el-Qurna, 18th Dynasty Egypt, ca 1550–1298 BC.

Figure 4.12 Pathway for the production of ethanol.

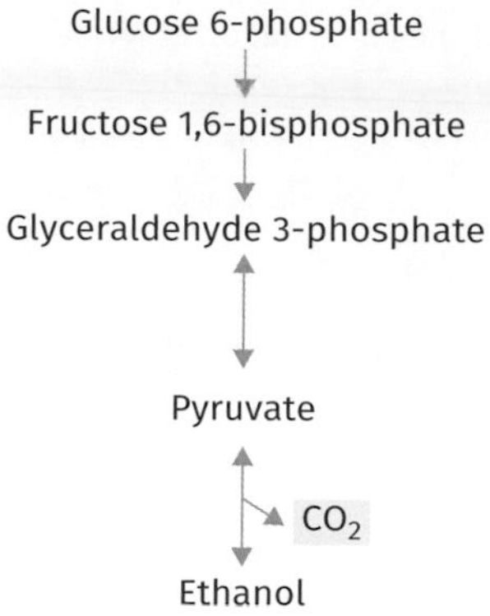

Figure 4.13 The tricarboxylic acid cycle with the glyoxylate shunt.

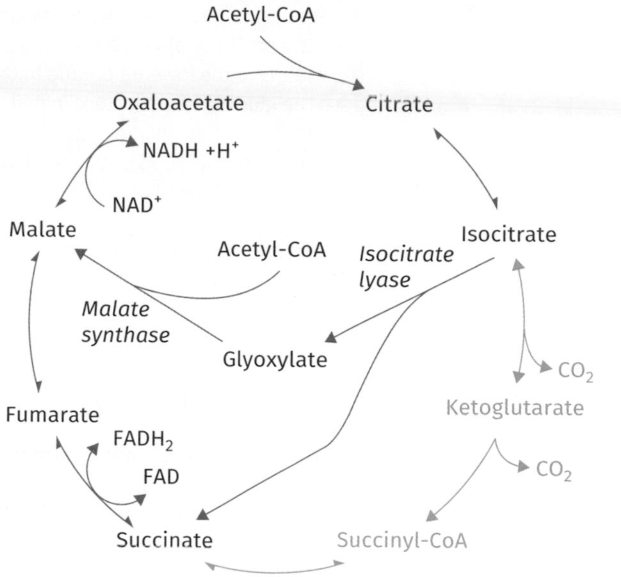

amylases break down the starch into glucose, sucrose, galactose, raffinose, and maltose. Baker's yeast ferments sugars in the dough into carbon dioxide, ethanol, and secondary metabolites (e.g. acetic acid, glycerol, succinic acid). This pathway (also see the Biochemistry primer in this series) is illustrated in Figure 4.12. Almost 1.5 litres of carbon dioxide can be produced from 100 g flour.

The succinic acid causes the pH in the dough to decrease. Succinic acid is produced via the oxidative pathway of the tricarboxylic acid cycle and the glyoxylate shunt, as shown in Figure 4.13.

The same principle for making bread is used around the world with flours of varying nutritional value (e.g. wheat, rye, spelt, corn), water, and sometimes salt. Many breads are unleavened. Chemical leavening (as opposed to yeast-based leavening) can be used, too, where natural acid is added with bicarbonate of soda to produce carbon dioxide.

Preparation of the aromatic and healthy sourdough bread starts with a combination of yeast species (e.g. *Saccharomyces cerevisiae*, *Saccharomyces exiguous* and *Candida* sp.) and heterofermentative lactic acid bacteria (e.g. *Lactobacillus brevis*). The starter, termed a sourdough, can be made at home by incubating flour and water at room temperature overnight, then adding fresh flour and water for a few days until the mixture bubbles soon after feeding and has a fresh, sour smell.

For industrial production the starter cultures are optimized to achieve highest bread quality and productivity.

Key points

- Microbial activity prolongs shelf life and changes food consistency and flavour.
- The flavour of a given food depends on the microbes and the specific incubation conditions.

- Fermented vegetable foods are produced using lactic acid bacteria.
- Vinegar is produced by acetic acid bacteria.
- The shelf life of cured meats is extended due to the combination of nitrite, salt, and the lowered pH due to lactic acid bacteria.
- The glucose of the lactose in milk is fermented to lactic acid, lowering the pH which curdles the milk.
- The microbial food web in food adds flavours.
- The open texture of bread is caused by the release of carbon dioxide by the yeast *Saccharomyces cerevisiae*.
- Sourdough is a combination of yeast species and heterofermentative lactic acid bacteria.

4.3 Microbes used to make alcoholic beverages

Microorganisms not only contribute to our food, but also to the drinks we consume—ranging from our breakfast beverages to wake up in the morning to alcoholic drinks for celebrations.

Alcoholic beverages

Alcoholic beverages have been regularly consumed for thousands of years—for example, bread-beer in ancient Sumer—because water from natural sources was not usually safe to drink until comparatively recently due to contamination with animal and human waste.

Just as yeasts use sugars in flour to generate the carbon dioxide that is so important in the context of bread-making, beverages with low alcohol content (produced by microorganisms) can start from various sources of sugars: grapes (wine), apples (cider), rice (sake), cereals (beer). The products are often distilled subsequently to produce spirits (e.g. whiskey from 'beer'), or yeast is added to bottles directly for further fermentation when making champagne. Table 4.5 shows different types of alcoholic beverages and their country of origin.

Table 4.5 Types of alcoholic beverages and their country of origin

Substrate	Non-distilled beverage	Product of alcoholic fermentation distilled to form	Location
Apples	Cider	Cider brandy	Northern Europe and N. America
Cacti/succulents	Pulque	Tequila	Mexico, Central America
Grapes	Wine	Cognac	S. Europe, N. and S. America, Australia, New Zealand
Pears	Perry	Pear brandy	N. Europe
Honey	Mead		UK
Sugar Cane		Rum	West Indies
Barley and other cereals	Beer	Whiskey	UK mostly
Rice	Sake	Shochu	Japan

Beer production

Beer brewing starts with malted grains prepared from germinated barley seeds. Brewer's yeast cannot produce the enzyme amylase, so the starch within the grain cannot be used until the seed has germinated, which releases hydrolytic enzymes such as amylases and proteases. After germination, the grains are kilned: first dried at 50–60°C and then cured at 80–110°C. This produces the malt. The malt is cooked as a mash with water, and natural amylases and proteases convert starch to sugar as a necessary preparation before brewer's yeast is eventually added. The resulting liquid is separated/filtered, yielding the wort. Female hops flowers can then be added for their flavour and antimicrobial properties. The mixture is boiled for several hours to sterilize and coagulate the proteins; it is then filtered and cooled.

The actual fermentation then takes place either using top- or bottom-dwelling yeast. *Saccharomyces cerevisiae* is a top-dwelling yeast, which after initial suspension in the liquid gets carried to the top by carbon dioxide. The process takes 5–7 days at 14–23°C. The bottom-dwelling yeasts (e.g. *S. carlsbergensis*) settle during the longer fermentation (8–14 days) at lower temperatures (6–12°C). The beer is then cooled, matured, and filtered. Sometimes a secondary fermentation is conducted to re-introduce more carbon dioxide into the beer before it gets packaged.

Beer can be brewed at various scales at home, in artisanal microbreweries, or large industrial facilities (see Figure 4.14, Figure 4.15, and Figure 4.16).

Wine production

Wine typically starts from grapes (but also other fruits), which are crushed to obtain the juice (called must). Cider production from apple juice or perry from pear juice follows a similar process to wine-making.

Fermentation can be started with natural yeasts (present on the surface of the grapes) or specific yeasts can be added as a starter culture (e.g. *Saccharomyces ellipsoideus*). Wine yeasts, *Saccharomyces cerevisiae, S. chevalieri, S. bayanus, S. italicus* and *S. uvarum*, are characterized by high ethanol tolerance and

Figure 4.14 Microbrewery.

Colin Woods/Shutterstock.com

Figure 4.15 Industrial-scale brewery.

DedMityay/Shutterstock.com

Figure 4.16 An example of brewing equipment for use at home.

underworld/Shutterstock.com

fermentation velocity. Production scales can range from a few litres at home to 200 litre casks and up to 200,000 litre tanks in industrial production. Red wine, for instance, is fermented for 3–5 days, then another 1–2 weeks until racking (separation of the sediment) takes place, followed by maturation, clarification, and further final maturation for many years.

Not just yeasts, but filamentous fungi play a role in wine production. The pathogen *Botrytis cinerea* causes grapes to dehydrate (Noble Rot), which concentrates the sugars and gives a sweeter flavour. This is somewhat comparable to ice wine where dehydration is caused by harvesting grapes very late, once they have been frozen on the vine.

Other alcoholic beverages

Kombucha is a fizzy, sweetened black or green tea following the fermentation by bacteria and yeasts (*Saccharomyces* and other species). *Gluconacetobacter xylinus* is always part of the bacterial population and oxidizes the alcohols produced by yeasts to acetic and other acids. This 'symbiotic culture of bacteria and yeast' is also known by the acronym SCOBY.

Non-alcoholic beverages

The microbial input into beverages without alcohol is not so much in the liquid end product, but in the process that matures the main basic ingredient. Cocoa and coffee beans are processed using natural fermentations (filamentous fungi, yeasts, acetic acid bacteria, and lactic acid bacteria), followed by drying, roasting, and grinding. Black tea leaves are withered and then rolled, enzymatically oxidized, and dried.

Coffee production

There are three main fermentation routes for coffee production. The most common one is low fermentation, a wet process during which the coffee beans (*Coffea* sp.) are washed after their skins have been removed, fermented in cement tanks, and then washed again. The remaining mucilage is then dried. This process leads to complex flavours.

For medium fermentation the deskinned beans are semi-washed and fermented. The mucilage continues to ferment on the bean while the sun is drying it all out, resulting in a sweet flavour. High fermentation is a dry process leading to coffee with the most complex flavours. The ripened beans are spread out as shown in Figure 4.17 and fermentation occurs in the individual beans.

Microorganisms used for the fermentation degrade the mucilage (e.g. pectinases in *Erwinia carotovora*), inhibit fungal growth (thus preventing contamination with mycotoxin), and produce the flavours. Specific starter cultures, mainly yeasts such as *Saccharomyces cerevisiae, Candida parapsilosis, Pichia guilliermondii, Saccharomyces apiculatus, Pichia fermentans, Candida bulgaricus*, and lactic acid bacteria, some *Enterobacteriaceae* and *Bacillus* sp., allow us to control the coffee quality.

Cocoa production

For cocoa production, the beans (*Theobroma cacao*) are fermented within their pods for seven days to allow microbial pectinases to degrade the pulp. The composition of the starter culture and the succession of microbial populations that are present is crucial for the development of the cocoa aroma. During the first 24 hours, yeasts are prevalent (with *Saccharomyces cerevisiae* being most abundant); lactic acid bacteria (*Lactobacillus fermentum, Lactobacillus plantarum*) then briefly dominate. Once the pulp vanishes and more oxygen is available, acetic acid bacteria (*Acetobacter tropicalis*) become prevalent. Over time, substrates such as sucrose, glucose, fructose, and citric acid are used up and products such as ethanol, lactic acid, and acetic acid are formed, as well as other

Figure 4.17 **Coffee bean ripening.**

Thanisnan Sukprasert/Shutterstock.com

volatile compounds which enhance the flavours. The temperature rises to 50°C which accelerates chemical reactions, leading to the curing of the beans. Then spore-forming bacteria and filamentous fungi can be observed.

The production of tea is somewhat different and almost resembles composting. Tea leaves (*Camellia sinensis*) are steamed, spread out, and sprinkled with water to cool. Fermentation starts if the leaves remain piled up. Caffeine content in tea leaves increases following fermentation. In Pu-erh tea *Aspergillus niger* dominates, but other fungi such as *Penicillium*, *Rhizopus*, and *Saccharomyces* can be isolated, as well as the bacteria *Actinoplanes* and *Streptomyces*.

Key points

- Beer brewing is based on malted grains prepared from germinated barley seeds. The fermentation uses top- or bottom-dwelling yeast.
- Wine production uses natural yeasts or specific yeasts added as starter culture. The microorganisms naturally occurring on grapes vary regionally and thus do the characteristic local flavours.
- There are three main fermentation routes for coffee production. The most common one is low fermentation/a wet process.
- The composition of microbial populations is crucial for the development of the cocoa aroma.
- The production of tea almost resembles composting.

4.4 Microbes as food: single-cell protein

Single-Cell Protein (SCP) is produced from the biomass of microbes, including single cell and filamentous fungi, bacteria, and algae. During the two World Wars there was a shortage of meat and fish protein. To boost protein supplies, interest in using microbial biomass as a source protein for both human and animal consumption grew. Strains of *Saccharomyces cerevisiae* and *Candida utilis* were grown to produce a product called Torula yeast, which was used to overcome protein shortages. The population explosion after the war made the availability of protein an even more urgent issue, and there were a number of different microbial protein production ventures, especially in the 1960s and 1970s.

SCP is a misleading term, as it's not a pure protein, but is in fact a whole microbial biomass which also contains carbohydrate, lipid, nucleic acids, minerals, and vitamins. There are distinct advantages in using microbial protein over plant and animal protein:

- microbes grow quickly and produce a lot of biomass;
- microbial cells have a high protein content (30–80 percent on a dry weight basis);
- production of microbial biomass occupies little land area when compared with plant crops and animal production;
- production of microbial biomass can occur all year round;
- microbes have an ability to use a wide range of substrates for growth, including agricultural waste products, e.g. soybean meal, which means that waste materials can be recycled;
- different strains of the microbe of interest can be selected and they can also undergo genetic modification;
- the product quality can be easily controlled.

Given all these advantages, why don't we all eat microbial protein? Well, there are some issues of safety and acceptability. One major issue with SCP is elevated nucleic acid levels, especially RNA. This high RNA is due to high growth rates of the microbe used for the biomass production and high protein content. Unfortunately, the digestion of RNA leads to the generation of purine compounds, resulting in a build-up of excess uric acid which can crystallize in the joints to give gout-like symptoms or kidney stones. The RNA has to be reduced to acceptable levels and this is usually done with heat (see the section on Quorn™ production a little later in this chapter). This step adds to the cost of SCP production, as well as affecting flavour, texture, and nutritional value.

Whilst most people are happy to eat mushrooms, the thought of eating bacteria and mould is not palatable to some people, although the growing market in probiotic products and the market in live yoghurts has made the idea of having bacteria as part of the diet more acceptable. SCP products may also have odd colouring and flavours. However, the manufacturers of Quorn™ have overcome consumers' reluctance to a large extent by promoting its health benefits.

What makes a good microbe for SCP production?

Various considerations need to be taken into account when determining whether a given microbe will be suitable for SCP production. These include the following:

- The microbe must have a high growth rate so that sufficient biomass is produced per g of substrate to make the process viable from a commercial perspective.
- The microbe should not require elevated temperatures for growth, as this will add to the cost of the production.
- The oxygen requirements of the microbe need considering: aerobic organisms will need oxygen to be supplied to the reactor.
- The microbe also needs to have genetic stability: it is undesirable for mutant strains to build up in the culture.
- The product needs to be recovered easily and it helps if the product can be readily shaped into familiar food forms.

Table 4.6 shows some of the advantages and disadvantages of different microbes as single-cell protein sources.

Types of cultivation methods

We discussed different types of cultivation method in Chapter 1, so this section is a quick reminder of submerged and solid state fermentation and how they relate to SCP.

In submerged fermentation the culture medium is always liquid and the reactor is operated continuously. Aeration is important, so that sufficient oxygen is available for aerobic respiration. Heat is also generated which has to be

Table 4.6 Microbes for SCP production: advantages and disadvantages

Microbe	Protein percentage	Nucleic acid percentage	Advantages	Disadvantages
Bacteria	50–85	10–16	Rapid growth rate Can grow on a variety of different substrates, high protein content	High levels of RNA
Algae	45–65	4–6	Simple cultivation, can photosynthesize, fast growth, and high protein content	No real disadvantages
Filamentous fungi	30–55	3–10	High amount of biomass produced, can grow on a variety of different substrates	Some fungi can produce very toxic compounds so screening is essential
Yeasts (single-cell fungi)	45–55	5–12	Rapid growth rate	More complex cultivation required

removed. The SCP product can be easily recovered from the culture by filtration or centrifugation. Submerged fermentation is good for the cultivation of yeasts and bacteria.

Solid-state fermentation uses a solid culture medium, e.g. rice or wheat bran compost. This type of fermentation does not require a reactor. Instead, cultivation can be done, for example, in trays in a room. Mushrooms are produced in this way and the product can be easily removed from the solid substrate.

Costs of SCP production: the carbon substrate

The overall cost of SCP is dependent on many variables, but one of the major costs of any SCP process is the carbon substrate, which can represent as much as 50–60 percent of the total process. Renewable resources are desirable, particularly agricultural, dairy, and wood-processing wastes. However, some substrates require more pre-treatment than others. The more reduced the substrate (e.g. oils and hydrocarbons) the better the biomass yield of the microbe. Despite being more expensive, the advantage is that they don't need any pre-treatment.

Oxidized substrates such as cellulosic wastes are cheaper, but they have to be pre-treated to release more readily utilizable carbohydrate units, such as glucose. The cost of the pre-treatment adds to the cost of the SCP production, and they are not converted into microbial biomass as efficiently.

Table 4.7 shows some of the carbon substrates which are used for the production of microbial biomass.

Table 4.7 Carbon substrates used for the production of microbial biomass

Carbon substrate	Microorganism
Carbon dioxide	Spirulina Chlorella
Liquid hydrocarbons	*Saccharomyces lipolytica* *Candida tropicalis*
Methane	*Methylomonas methanica* *Methylococcus capsulatus*
Methanol	*Candida boidinii*
Ethanol	*Candida utilis*
Glucose	*Fusarium venenatum*
Inulin	*Candida* spp.
Molasses	*Candida utilis* *Saccharomyces cerevisiae*
Spent sulphite waste	*Paecilomyces variotii*
Whey	*Kluveromyces marxianus* *Penicillin cyclopium*
Lignocellulose waste	*Agaricus bisporus* *Cellulomonas* spp.

A successful SCP production: Quorn™

Despite the failure of some SCP production, for example Pruteen—an animal feed manufactured by ICI—one product has been a massive success. In 1991, Marlow Foods (a subsidiary of ICI) isolated a species of *Fusarium graminearum*, a filamentous fungus, from soil in Buckinghamshire. The name is now familiar to us as Quorn™. Table 4.8 shows the comparison between Quorn and more traditional sources of animal protein. Quorn™ is marketed as a healthy food and has 12 percent protein with no animal fat or cholesterol, and reduces the fats in the blood stream. The chitinous cell wall of the fungus also acts as a source of dietary fibre, although it cannot be digested by humans, and there is a high vitamin B content.

The annual sales of Quorn™ are now in excess of £150 million in the UK and 1,000 tonnes of Quorn™ are produced each year in the 40 m^3 airlift fermenter in the north east of England, which was originally designed to make Pruteen. Remember from Chapter 1 that an airlift fermenter does not have a stirrer, but relies on air pumped in from below the vessel to mix the cell culture. The fermenter is operated continuously at 30°C and pH 6.0. The fungus, *Fusarium venenatum* (which was renamed from *Fusarium graminearum*) is grown on food-grade glucose syrup as the carbon source and ammonia is used both as the nitrogen source and also to control the pH. Growth of the fungus is supplemented with biotin and mineral salts. The oxygen levels have to be carefully controlled: if the *F. venenatum* grows anaerobically, unpleasant flavours can be produced in the final product.

Quorn has a fibrous structure in its mycelium which resembles meat; this feature can be used to mimic burgers, sausages, and minced meat. However, when grown in prolonged culture, highly branched forms of the fungus can occur after approximately 100–1200 hours. When detected, the fermentation has to be shut down as the texture of the final product loses its fibrous nature.

The fungal mycelium is harvested by vacuum filtration. The 'filter cake' produced is a matt of intertwined fungal hyphae. A range of Quorn™ products can be seen in Figure 4.18.

The final biomass contains 10 percent RNA which is too high for humans to eat, so RNA levels have to be reduced by thermal shock at 64°C for 30 minutes.

Table 4.8 A comparison of Quorn™ to animal protein sources

	Cheddar cheese	Raw chicken	Raw beef	Fresh cod	Quorn™
Protein ($g\ 100g^{-1}$)	26.0	20.5	20.3	17.4	12.2
Dietary fibre ($g\ 100g^{-1}$)	0	0	0	0	5.1
Total fats ($g\ 100g^{-1}$)	33.5	4.3	4.6	0.7	2.9
Fats ratio (poly-unsaturated: saturated)	0.2	0.5	0.1	2.2	2.5
Cholesterol ($mg\ 100g^{-1}$)	70	69	59	50	0
Energy ($kJ\ 100g^{-1}$)	1697	506	514	318	334

Figure 4.18 **The different products from the mycoprotein Quorn.**™

Philip Reeve/Shutterstock.com

This kills the fungus and activates RNases which break down the RNA into nucleotides, which then diffuse out of the cells. By using this method RNA is reduced to 2 percent of the original levels. The company which now owns the Quorn™ brand is set to expand production with a recently announced £150 million investment, and sales of Quorn™ are projected to steadily increase.

Dihé

The production of SCP does not have to be highly technical or use specialized equipment. This is clearly illustrated in Figure 4.19. The cyanobacterium *Spirulina platensis* is used to make Dihé in the African regions of Kanem and Lake Chad. In Africa, *S. platensis* is taken from Lake Chad, and poured into a depression created in the ground. When dried, the *S. platensis* is cut into chunks and sold. *Spirulina platensis* is also sold as a health food supplement.

Figure 4.19 **The production of Dihé in Africa.**

©FAO/Marzio Marzot

Mushroom production

Mushroom production is a substantial industry. The major cultivated mushroom is the button mushroom *Agaricus bisporus* and is worth approx £1 billion per annum. Other commercial species include the oyster mushroom *Pleurotus ostreatus*, and the shitake mushroom *Lentinula edodes*, which is grown on logs.

Mushroom production is a very efficient use of plant waste material. The product is directly edible and harvesting is easy as production uses solid-state fermentation. Compared with animal protein production, there is good protein conversion efficiency when comparing the amount of land needed for mushroom production and the time taken for growth.

Mushrooms are grown in a solid-state fermentation, a fermentation process which does not have water (or any other liquid). It can be carried out in small spaces, as shown in Figure 4.20. The solid substrate used in the fermentation can be made from a variety of different materials depending on the country of production; typically, this is composting straw, manure, and fertilizer.

The production of the compost has four phases:

1. The components are mixed together and the mesophilic microorganisms (those that grow between 20°C and 45°C) develop. The carbohydrates and the proteins within the mix are converted to heat and ammonia. As the temperature rises the thermophilic microorganisms develop, which softens the straw.
2. The compost is conditioned for eight hours at 56°C and then kept at 45°C. The microorganisms at this stage are *Actinomycete* bacteria, and also the thermophilic fungus *Humicola insolens*. The inoculum preparation of *A. bisporus* starts with the growth of the spawn on sterilized cereal grains, e.g. millet or rye. The spawn is added to the compost at the end of Phase 2. The mycelium grows through the compost for between 12–16 days.

Figure 4.20 Mushroom production in solid state fermentation using horizontal trays.

Kartinkin77/Shutterstock.com

3. Temperatures fall to 20–25°C.
4. The fruiting bodies of *A. bisporus* then start to appear and can be harvested after 18–21 days. The fruiting bodies are picked, but they continue to appear (flush) in cycles of 7–8 days.

Key points

- Microbial biomass, known as Single-Cell Protein (SCP), can be used as a source food for animals and humans.
- Different microbes can be used to make biomass for consumption, but there are issues with safety and acceptability.
- The major cost of a SCP process is the carbon substrate used in the fermentation.
- Many SCP processes have failed due to the cost of production and limited demand.
- Quorn™ is an example of a successful SCP product.
- Mushroom production is an example of SCP grown in solid-state fermentation.
- Production of SCP need not be highly technical, e.g. production of Dihé.

4.5 Microbes producing food supplements

In section 4.1 we covered how microbes are used to make components which are then used to manufacture different foods and drinks. This section will cover the products from microbes which are used as food supplements. Food supplements include pigments, flavourings, and the growing probiotic market; however, we will first consider the production of vitamins.

Vitamins

We covered the production of vitamins as a fine chemical in Chapter 3, but this section examines their more specific role as a food supplement. As we are becoming more health conscious the market for vitamin supplements has grown dramatically. In the UK we spent over £400 million on vitamins in 2015, and it's estimated that 50 percent of us take them on a daily basis. Vitamins used to be made from both plant and animal biomass, but microbes are now used as the major source. For example, strains of the single cell fungus *Saccharomyces cerevisiae* are used to make a range of B vitamins. Take a look at Table 4.9 for more examples.

Vitamin C has been in the human consciousness for thousands of years; however, it wasn't known by its name or its chemical designation—L-ascorbic acid, but by the disease caused by its lack in the diet: scurvy. The symptoms were first described in the famous Ebers papyrus in 1550 BC. Vitamin C is a cofactor in collagen synthesis and a lack of it in the diet results in collagen instability and thus loss of teeth. Vitamin C also acts as an antioxidant and is therefore used in the prevention of infectious diseases such as flu, against ageing, and cancer. Back in the 1960s, Linus Carl Pauling (double Nobel Laureate and biochemist) promoted the use of high doses of vitamin C to prevent diseases. The antioxidative capacity of the compound is also used by the food and beverage industry, consuming 25 percent and 15 percent respectively, of all

Table 4.9 Microbial production of vitamins and their uses

Product	Producing organism	Use
Vitamin C	Biotransformation process	Nutritional supplement
B vitamins	Yeasts—*Saccharomyces cerevisiae*	Nutritional supplement
B_{12}	*Pseudomonas denitrificans* and *Propionibacterium shermanii*	Nutritional supplement
β-carotene	*Phycomyces blakeleeanus*	Precursor of vitamin A; colouring agent for margarine and baked goods
Ergosterol	*Saccharomyces cerevisiae*	Source of vitamin D
Riboflavin	*Ashybya gossypii*	Nutritional supplement

Table 4.10 Examples of E numbers based on vitamin C

E number	Chemical form
E300	Free ascorbic acid
E301–303	Ascorbate salts of sodium, calcium, or potassium, respectively
E304	Fatty esters
E315	Erythorbic acid, a stereoisomer

vitamin C produced, to extend shelf-life and protect colour, flavour, and nutrient content. Various E numbers (the different chemical forms of vitamin C) are used as food additives as listed in Table 4.10.

The production of vitamin C is covered in Case study 4.1.

Case study 4.1
The production of vitamin C

Animal and plants synthesize vitamin C (L-ascorbic acid) starting from D-glucose, whereas fungi such as *Zygomycetes*, *Ascomycetes*, or *Basidiomycetes* start from D-arabinose. Industrial production started with extraction of the vitamin from fruit and the first chemical process dates back to 1933. More than 100,000 tons of L-ascorbic acid and its derivatives are produced each year making vitamin C the largest vitamin amount in production. The Reichstein–Grussner process (from 1934) is based on chemically hydrogenating D-glucose to D-sorbitol, which is then microbially oxidized by *Gluconobacter* sp. strains to L-sorbose, followed by further chemical conversions. This process is still used today with a productivity of more than 50 percent. A two-stage microbial process was introduced in the late 1960s in China. Following the L-sorbose production by *Gluconobacter* sp., 2-keto-gluconic acid is produced via L-sorbosone by *Ketogulonicigenium* sp., which then again is chemically converted to the vitamin. This procedure has less steps in total and is therefore cheaper, and needs less energy and chemicals.

Pigments

We are often told by chefs and others in the restaurant industry that we 'eat with our eyes', thus it follows that the colour of our food impacts upon our perceptions of it. Colour can also be an indication of freshness, and therefore safety. Manufacturers of processed food products are looking to replace synthetic pigments with natural ones, as natural products are perceived to be healthier and safer; they also tend to be more light stable, and can tolerate changes in heat and pH. Thus, pigments produced by microorganisms are seen to be a viable alternative to chemical synthesis. Important natural pigments include carotenoids, tetrapirroles, flavonoids, and xantophylls; more examples can be found in Table 4.11. It may surprise you that microbial pigments have been used for thousands of years in China. Records from the first century describe the use of Anka, a red pigment from red mould rice. It is used as a colorant and spice. One of the most successful industrial productions is the pigment β-carotene, produced by the filamentous fungus *Blakeslea trispora*. Glucose and corn steep liquor are used as the carbon and nitrogen sources and the pigment is made in aerobic submerged batch fermentation. The β-carotene is removed from the fungal biomass using ethyl acetate and then crystallized.

Table 4.11 The use of microbial pigments in the food industry

Pigment	Colour	Producing organism	Use in the food industry
Riboflavin	Yellow	Filamentous fungus—*Ashbya gossypii* Yeast—*Candida famata* Bacteria—*Bacillus subtilis*	Baby food, breakfast cereals, pasta, sauces, processed cheese, fruit drinks
Monascus pigments, e.g. ankaflavin	Yellow, red, and purple	Filamentous fungus *Monascus* spp.	Food colouring in processed meats and fish paste
Beta-carotene	Red	Filamentous fungus—*Phycomyces* spp. and *Mucor circinelloides*	Margarine
Canthaxanthin		Bacteria—*Bradyrhizobium*	Aquaculture where it is a flesh pigmentor
Carotenoids	Yellow-orange, red	Bacteria—*Streptomyces* and *Agrobacterium*	Margarine
Astaxanthin	Red	Yeast—*Xanthophyllomyces dendrorhous*	Feed and aquaculture where it is a flesh pigmentor
Indigoidine	Blue	*Corynebacterium insidiosum*	
Phycocyanin	Blue	Cyanobacteria	
Violacein	Purple	*Chromobacterium violaceum*	Medicine, cosmetics, foods, and textiles
Zeaxanthin	Yellow	*Flavobacterium* spp.	Poultry feed to strengthen the yellow colour of the skin

Artificial flavours

Microbes, as we have seen in earlier section of this chapter, are incredibly important in the development of flavour of a variety of different foods and beverages, e.g. cheese, soy sauce, beer, and wine. Microbes also have a potential role in the production of artificial flavours, although the biggest challenge is that yields tend to be very poor, as flavour compounds tend to be inhibitory, which does not make the processes commercially viable.

A number of flavours and their producing organisms are outlined in Table 4.12. We will then examine the potential production of vanillin in more detail in Case study 4.2.

Table 4.12 Examples of flavour compounds and their potential producer organisms

Flavour compound	Flavour	Producing organism(s)
Diacetyl	Buttery	Lactic acid bacteria, including *Lactococcus lactis* and *Streptococcus thermophilus*
Lactones (cyclic esters)	Fruity, coconut-like, creamy	Filametous fungus *Trichoderma virdie*, and the yeast *Candida tropicalis*
Esters	Fruity aromas	The yeasts *Hanseniaspora guilliermondii, Pichia anomala*. The bacterium *Lactococcus lactis*
Pyrazine	Roast flavours	Bacteria such as *Corynebacterium glutamicum*
2-phenylethanol	Rose flavour	Yeasts such as *Hansenula anomola* and *Saccharomyces cerevisiae*
Benzaldehyde	Cherry and fruit flavour	The bacterium *Pseudomonas putida*, and the white rot fungus *Trametes suaveolans*

Case study 4.2
Vanilla production

Vanillin is an aromatic aldehyde ($C_8H_8O_3$) and it is an important flavouring in many foods such as cakes, ice cream, and soft drinks, but it also can be used as a preservative. Vanillin is found naturally in the bean or pod of the orchid *Vanilla planifolia*, see Figure CS 4.21.

Figure CS 4.21 **Vanilla planifolia.**

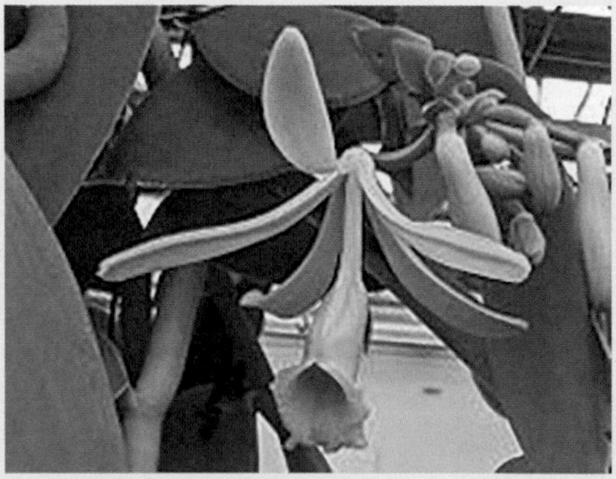

guentermanaus/Shutterstock.com

The main producing countries are Indonesia (3,700 tonnes in 2008), Madagascar (2,800 tonnes in 2008), and China (1,400 tonnes in 2008). However less than 1 percent of the vanilla we use in food comes from a natural source as it is expensive; most is produced more cheaply by chemical synthesis. Vanillin is an aromatic aldehyde, 3-methoxy-4-hydroxybenzaldehyde, it is a white solid and its extraction from the vanilla pod is long and expensive. Vanillin can be made chemically from the controlled oxidation of lignin from wood waste. Other methods include the biotransformation of caffeic acid from *Capsicum frutescens*, shown in Figure CS 4.22.

Microbes also have potential in making vanillin. Ferulic acid (see Figure CS 4.22) is a phenolic compound from lignin degradation and it is a vanillin precursor. Several bacteria can use ferulic acid as a sole carbon source to produce vanillin, examples include the Gram-negative bacterium *Pseudomonas fluorescens* and also Gram-positive bacteria *Amycolatopsis* spp. and *Streptomyces setonii*. Getting the conditions right for high levels of vanillin production is difficult, so the use of recombinant DNA technology here could be very useful. Research has shown that cloning the genes for feruloyl-CoA synthase and feruloyl-CoA hydratase into suitable plasmid vectors and transformation in an *E. coli* host allows for the production of vanillin from ferulic acid.

Figure CS 4.22 Biotransformation of caffeic acid.

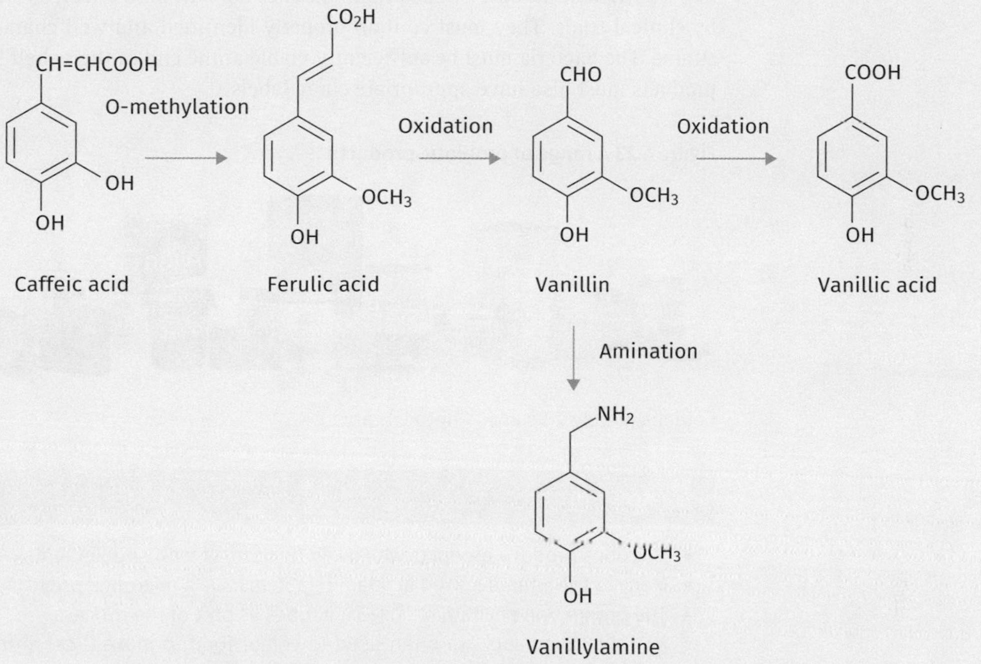

Probiotics

Probiotics are 'friendly' bacteria. The popularity of probiotic products has increased dramatically over the last decade and as you can see from Figure 4.23 there is now a whole range of products available from drinks and chocolate to tablets. Probiotics are often portrayed in advertisements as 'helping to speed digestive transit' and we spend millions on them in the hope that they will improve our health. The market worldwide for probiotics is thought to be in excess of 44 billion USD. The definition of a probiotic is 'live microorganisms that when administered in adequate amounts confer a health benefit on the host' (World Health Organization). Probiotics must be able to withstand and survive the effect of gastric acid, biliary secretions, and pancreatic secretions in order to reach the small and large intestines alive.

Probiotics are not genera or species; types of bacteria don't become a probiotic until shown by evidence to confer a specific health benefit. Typical organisms associated with probiotic products include *Lactobacillus* spp. and *Bifidobacterium* spp., as well as some yeasts and Gram-negative bacteria. There have been many trials conducted, and there are numerous reviews and meta-analyses available in the scientific literature, which suggest that some probiotic organisms can help with the treatment of diarrhoea associated with antibiotic treatment, boost the immune system, and help alleviate the symptoms of Crohn's disease which is a condition of the gut. However, the EU Food Standards Agency has reviewed probiotic health claims and decided the health benefits claimed are not justified by the independence of the evidence presented by companies marketing probiotics.

In order for probiotic products to have widespread credibility, they must contain a sufficient number of bacteria in order to see a desired effect, as suggested by clinical trials. They must contain properly identified and well characterized strains. The bacteria must be sufficiently viable at the end of their shelf life. The products must also have appropriate claim labels.

Figure 4.23 A range of probiotic products.

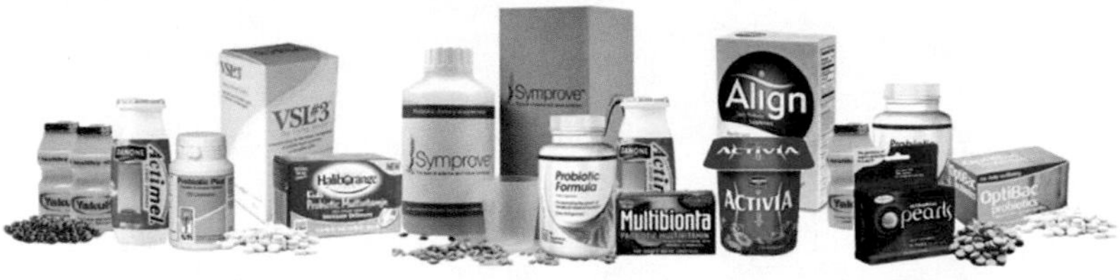

Cordelia Molloy / Science Photo Library

Key points

- Microbes produce products which can be used as food supplements.
- Many of the vitamins used in vitamin tablets have a microbial origin.
- The production of vitamin C uses microbes as part of the process.
- Microbial pigments are often used to colour food to make it seem fresher and more palatable.
- Microbes are important in the development of food flavour. Flavourings of microbial origin can be used in food and drink manufacture.
- The probiotic industry is large, and growing. There are now numerous food-based and supplement products which are designed to be taken regularly to improve health.

Chapter Summary

- Humans have had a long and rich history of using microbes for food and beverage production.
- Microbes can be used to make components for food production, these include amino acids, enzymes, oils, and polysaccharides.
- Microbes are used to make food directly; examples include tempeh, fermented meat products, cheese, and bread.

- Microbes are used to make alcoholic beverages such as beer and wine.
- Microbes can be used as food. These products come under the term 'single-cell protein'.
- The product Quorn™ is an example of a highly successful single-cell protein product.
- The production of single-cell protein does not have to be highly technical; examples of low-cost production include Dihé.
- Mushroom production is an example of a single-cell production which is carried out using solid-state fermentation.
- Microbes can be used to produce food supplements such as vitamins, pigments, and flavour enhancers.
- Microbial activity prolongs shelf-life and changes food consistency and flavour.
- The microbial food web in food adds flavours.
- Different microbes can be used to make biomass for consumption, but there are issues with safety and acceptability.

Further Reading

Kumar, R., Vikramachakravarthi, D., and Pal, P. (2014). 'Production and purification of glutamic acid: A critical review towards process intensification'. Chemical Engineering and Processing: Process Intensification 81: 59–71.

For a more in-depth look at the production of glutamic acid, read the review by Kumar, Vikramachakravarthi, and Pal. This review also covers some of the biochemical aspects of glutamic acid production.

https://theconversation.com/why-the-meat-industry-could-win-big-from-the-switch-to-veggie-lifestyles-112714

For an interesting conversation article on the meat industry developing meat-free protein, read the following article:

Tamang J. P., Watanabe, K., and Holzapfel W. H. (2016). 'Review: diversity of microorganisms in global fermented foods and beverages'. Frontiers in Microbiology 7: 377. doi: 10.3389/fmicb.2016.00377 https://www.frontiersin.org/articles/10.3389/fmicb.2018.01746/full

This article offers a detailed overview of fermented world foods and beverages.

Carroll A. L., Desai, S. H., and Atsumi, S. (2016). 'Microbial production of scent and flavor compounds'. Current Opinion in Biotechnology 37: 8–15. doi.org/10.1016/j.copbio.2015.09.003

For a detailed overview of scents and flavours see the following article:

Discussion Questions

4.1 Discuss the idea that single-cell protein could be the answer to the food crisis in the developing world.

4.2 Biotechnology is one of the biggest growth industries worldwide. Discuss.

4.3 Discuss how the connection of microorganisms with food and beverages can maintain our health, be beneficial or harmful to it.

4.4 Based on what you have learnt in this chapter, discuss the roles and responsibilities microbiologists can have in food and beverage production.

5 ENVIRONMENTAL BIOTECHNOLOGY

Learning Objectives

- To identify the role microbes play in wastewater treatment and anaerobic digestion;
- to describe the processes of wastewater treatment and anaerobic digestion;
- to recognize that microbes degrade hydrocarbons and play a role in cleaning up oil spills;
- to understand how microbial bioleaching can be used to recover metals from low-grade ore and minimize pollution in local environments;
- to describe how a biosensor functions and the use of microbes in these devices;
- to understand how bioscrubbing can be used to remove biodegradable contaminants from gas streams;
- to understand the use and risks involved in bioaugmentation, highlighted by real-life examples.

Environmental biotechnology harnesses natural processes for commercial applications. The International Society for Environmental Biotechnology defines it as 'the development, use and regulation of biological systems (i.e. microorganisms) for **bioremediation** of contaminated environments (land, air, water), and for environment friendly processes (green manufacturing technologies and sustainable development)' Bioremediation may occur naturally through microorganisms found in the polluted area, or be augmented by human activity, either by the addition of foreign microbial species or by altering the environment to facilitate the process of required local species enrichment. It is a major industry, with water treatment alone constituting a global market worth approximately 150 billion USD in 2015. With an increasing and rapidly urbanizing global population, the importance of biotechnology for minimizing the impact of human activity on the natural environment is increasingly important. This chapter will explore a range of topics in the area, including wastewater treatment, anaerobic digestion, hydrocarbon bioremediation, environmental monitoring, and metal leaching with a focus on the role of microorganisms in each process.

5.1 Wastewater treatment

Wastewater derives from households, livestock, and industry, and depending on the source, may contain high concentrations of faeces and urine, various organic compounds, toxic metals, and microbial pathogens. Bioremediation of wastewater involves reducing or removing these contaminants to concentrations where the water can be reused or released into the local ecosystem. Bacteria are the most numerous pathogens in wastewater; most common are *Salmonella* spp., *Vibrio cholerae*, and *Shigella* spp., although high numbers of these organisms are generally required to cause illness. The most frequent waterborne illnesses are caused by enteric viruses, which are shed in high numbers in faecal material. These virus particles, such as norovirus and rotavirus, are not removed by filtration. Eukaryotic microbes such as protozoa produce cysts and oocysts that can withstand harsh environments and are also chlorine resistant. Examples include *Giardia* and *Cryptosporidium*, which are some of the most common waterborne infections in the UK.

Looking at Figure 5.1 you will see a schematic of the various processes involved in household wastewater treatment which are explained in the following paragraph.

The first step involves transporting waste from the source to the treatment plant, predominantly via sewers, one of the most important developments in human history (see The bigger picture panel 5.1). Upon arrival at the plant, large solids are removed using screens, whilst smaller particles, termed sludge, are separated from the main water liquid body via sedimentation. This reduces the biological oxygen demand (BOD) by 30–40 percent. BOD is a measure of the amount of oxygen needed by microorganisms to oxidize the organic matter in the water, and is an indicator of pollution.

In the secondary phase, the wastewater is treated with both biological and physical processes. The object is to speed up the biological degradation using microbes and further reduce the BOD by 80–90 percent. The liquid is

Figure 5.1 A schematic diagram detailing the wastewater treatment process.

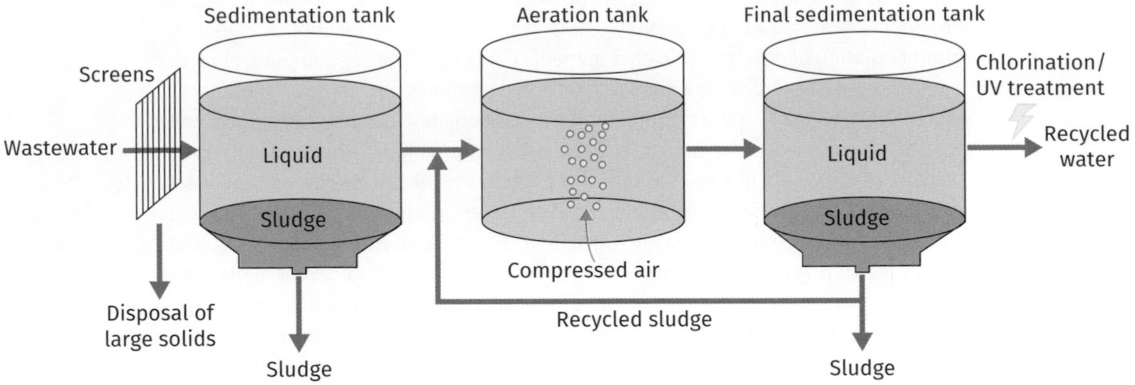

first processed through an aeration tank where aerobic oxidation of organic matter occurs, primarily by hundreds of different strains of microorganisms, predominantly β-*proteobacteria*, but also α-, ε- and γ-*proteobacteria*, and *Acinetobacter*, *Zoogloea*, and *Hyphomicrobium* bacterial species. Aerobic processes can involve using trickling filters containing beds of stones or moulded plastic, which the water trickles over. Microbial biofilms develop on the surface of the stones, creating complex communities of bacteria, fungi, and protozoa. Another secondary process is activated sludge treatment which is aerobic and needs continual agitation. The wastewater is bought into contact with a mixed microbial population in the form of flocs. These flocs are formed by the production of microbial polysaccharides by *Zoogloea*, *Acinetobacter*, *Flavobacterium*, *Paracoccus*, *Alcaligenes*, *Corynebacter*, and *Pseudomonas* species. The bacteria are prey for eukaryotic microbes such as ciliates and amoebas. Carbon dioxide and inorganic nutrients (i.e. PO_4^{2-}, NO_3^-, SO_4^{2-}) are released during this process.

Solids and biomass from this process are then separated from the liquid by sedimentation, with a proportion recycled through the aeration tank, while the remaining sludge is added to that previously collected. Depending on the source, the liquid is then disinfected via either ultraviolet light or chlorine before being used for agricultural, industrial or domestic use, or released into the environment. The sludge generated throughout this process is subsequently dewatered and subjected to anaerobic digestion (discussed in section 5.2), composted, used as fertilizer, or disposed of either in landfill or by incineration.

Key points

- Wastewater treatment can be used to treat water from a variety of sources containing different pollutants.
- A range of microorganisms degrade organic particles and the process is aided by aeration and mixing of cultures.
- Biological oxygen demand (BOD) is used to determine whether the water can be reused or released into the local environment.

The bigger picture panel 5.1

Sanitation, the greatest medical advance of the modern era

In 2007, *British Medical Journal* readers chose the introduction of clean water and sewage disposal as the most important medical milestone since 1840. Prior to the development of the sewage system, drinking water in cities was often taken from polluted rivers. This resulted in regular disease outbreaks, particularly of cholera. To solve this issue in London, an integrated modern system comprising over 20,000 kilometres of sewers was built that diverted waste to the Thames Estuary, downstream of the city. Similar systems were soon adopted in other industrialized nations. Poor sanitation and consequentially outbreaks of cholera and other infectious diseases, is still an issue in the developing world, exacerbated by rapidly growing urban populations.

What are possible solutions for sanitation in poor urban centres of the developing world?

Figure BP 5.1 (a) Dr John Snow, who demonstrated that cholera was borne in water contaminated with sewage; (b) workers in one of the main London sewers in the 19th century; (c) people walking along an open sewer in an African township.

5.2 Anaerobic digestion of solid waste

Anaerobic digestion (AD) is the process by which microorganisms break down biodegradable material in the absence of oxygen. Thousands of anaerobic digestion plants are currently operating worldwide, using landfill, agricultural residues (i.e. corn cobs, wood, straw), wastewater sludge, food waste, or manure as a feedstock. In theory, the process is relatively simple and is outlined in Figure 5.2 and the following paragraphs.

The feedstock is sealed in a tank or 'digester'. Naturally occurring microorganisms, typically a consortium of bacteria and archaea, digest the biomass at either **mesophilic** (25–45°C) or **thermophilic** (50–60°C) temperatures. In most cases, the major products are a methane-rich biogas, which is subsequently

Figure 5.2 A schematic diagram detailing the anaerobic digestate process.

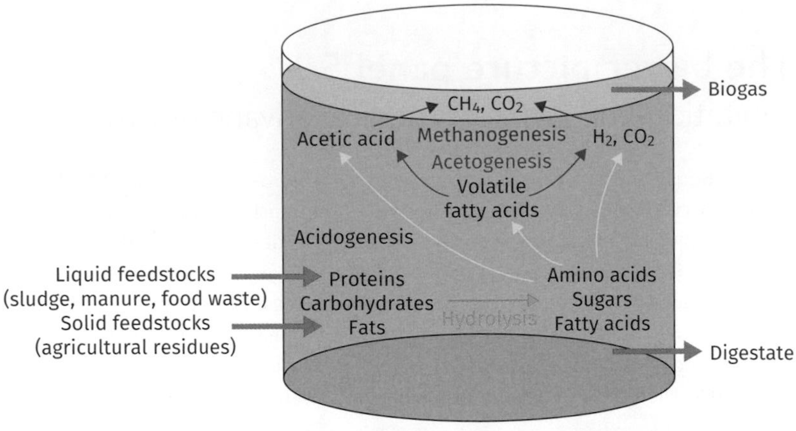

burnt to produce electricity, and a solid nutrient-rich material called digestate, which is used as fertilizer. However, optimizing the process to increase digestion and produce the highest amount of methane is critical for commercial success. Animal manure, sludge, and organic household waste are typically added to a digester as a liquid with solid concentrations <15 percent (liquid AD), whereas agricultural residues are added at higher concentrations (solid-state AD). Due to the high concentration of **lignocellulose** in agricultural residues these feedstocks are often pre-treated by a combination of heat/acid/steam (Chapter 1, section 1.2) to facilitate digestion by the microbial community. Solid-state AD utilizes smaller tanks and limited stirring which reduces heating and running costs but requires longer periods for digestion than liquid AD.

Conversion of organic material to methane occurs via four major enzymatic processes, each performed by a different group of microorganisms within the same tank.

1. The first process, **hydrolysis**, results in breakdown of large polymers such as carbohydrates, proteins, or lipids to their respective monomers, sugars, amino acids, and fatty acids. Extracellular enzymes secreted by hydrolytic bacteria, mostly *Cellulomonas, Clostridium, Bacillus, Thermomonospora, Ruminococcus, Baceriodes, Erwinia, Acetovibrio, Microbispora*, and *Streptomyces* species, catalyse these reactions. This is often the rate-limiting step, especially when polymers like lignin and cellulose are highly abundant.
2. The second process, fermentation, involves the conversion of sugars, amino acids, and fatty acids to H_2 and CO_2, acetate or other volatile fatty acids (e.g. propionate, isobutyrate, butyrate, valerate, isovalerate), catalysed by predominantly *Acetobacterium, Clostridium, Sporomusa, Butyribacterium, Lactobacillus*, and *Streptococcus* bacterial species, in addition to *Saccharomyces* yeast.
3. The third process, **acetogenesis**, results in the conversion of volatile fatty acids to H_2 and CO_2, and acetate, by acetogenic bacteria, or **acetogens**, which are predominantly *Acetobacterium, Clostridium, Sporomusa, Ruminococcus*, and *Eubacterium* species.

4. The fourth process, **methanogenesis**, involves the conversion of acetate and H_2/CO_2 generated from processes 2 and 3 to methane by methanogens, which are exclusively archaea.

The efficiency of the process is dependent on controlling environmental parameters such as temperature and pH, which should be maintained at ~pH 7, but can be affected by the relative concentrations of the intermediates, temperature, and substrate inputs. The pH can be altered by adding either alkaline agents such as lime and soda ash, or acids such as muriatic acid. Ideally, the same feedstock should be used without any substances that inhibit microbial growth or introduce novel microorganisms. This results in a stable microbial population and consistent methane production. Stability of the microbial population is generally possible with agricultural residues but more challenging with sludge, animal manure, or household waste, which is often more variable in its composition.

Key points

- Anaerobic digestion occurs via four main processes: hydrolysis, fermentation, acetogenesis, and methanogenesis.
- A consortium of bacteria, yeast, and archaea are involved.
- Temperature, pH, and feedstock inputs have to be tightly controlled for optimal methane production.

5.3 Hydrocarbon bioremediation

Hydrocarbon release from natural sources (e.g. seeps, biological production) and human activities (e.g. shipping, oil rigs and wells, vehicles) occurs in nearly all environments on Earth. Irregular events such as the grounding of the Exxon Valdez tanker in 1989 and the Deepwater Horizon oil rig explosion in 2010 (see The bigger picture panel 5.2) release large amounts of petroleum in a localized region, with devastating effects on the environment. Petroleum, as illustrated in Figure 5.3(a), is a complex mixture of alkanes, alkenes, aromatics, and more complex asphaltenes and resins. Each group contains compounds of varying sizes, which determine their physical properties. C1–C4 hydrocarbons, including compounds like methane and ethane, are gases. C5–C8 hydrocarbons such as pentane and benzene are water soluble. Therefore, their release into the environment is not visible. Hydrocarbons greater than C8 are either poorly water soluble or insoluble, and therefore are clearly visible as oil slicks. Many of these compounds are toxic and/or carcinogenic.

Certain microorganisms have evolved mechanisms to import and metabolize different hydrocarbons. Methane is released in large amounts by archaea inhabiting environments as diverse as wetlands, rice paddies, and the guts of ruminants. Consequently, methanotrophs, which include specialized α- and γ-proteobacteria species belonging to bacterial families such as Methylococcaceae and Methylocystaceae, have evolved to metabolize methane and similar compounds, and are widespread in different environments. An example of a methane metabolism pathway from Methylococcaceae is shown in Figure 5.3(b). In this pathway, methane is converted to methanol which in turn

The bigger picture panel 5.2
Cleaning up after oil spills

While hydrocarbon release into the environment is common, it is large spills such as the rupture of the tanker, the Exxon Valdez, off the coast of Alaska in 1989 and the Deepwater Horizon rig disaster in the Gulf of Mexico in 2010, which demonstrate the damage that releasing large amounts of oil into the environment can cause. Various efforts have been conducted to clean up the oil, including burning it, manually removing it, and adding dispersants (to break it apart) and fertilizers (to stimulate growth of hydrocarbon degrading bacteria). Despite these efforts, residual oil is still present in the Alaskan environment and populations of some animals have never recovered.

What is the fate of the soluble hydrocarbons following an oil spill?

Which hydrocarbons are going to be the most difficult to remove from the environment?

Figure BP 5.2 (a) Oil spilling from the Exxon Valdez tanker; (b) efforts to control the blaze following the explosion at the Deepwater Horizon oil rig.

(a) (b)

(a) Anchorage Daily News / Tribune News Service/ Getty Images (b) Uncredited/AP/Shutterstock

is converted to formaldehyde. Formaldehyde can be assimilated into biomass or oxidised to formate then CO_2 to generate reducing equivalents (NADH).

Larger hydrocarbons, predominantly alkanes and alkenes (C15–C38), are produced by cyanobacteria, algae, and plants, with considerable quantities subsequently released into the environment. As a consequence, microorganisms that metabolize these hydrocarbons, termed hydrocarbonoclastic, are widespread in marine, freshwater, and terrestrial environments. These include specialized bacterial, fungal, and algal species. Some, such as the bacterium *Alcanivorax borkumensis*, are obligate hydrocarbon-degrading organisms and unable to utilize most other carbon sources for growth and energy. Larger hydrocarbons, derived from either biological sources or oil spills, aggregate as oil droplets. Certain hydrocarbonoclastic microorganisms release dispersants, such as rhamnolipids, into the external environment to break apart these droplets. Hydrocarbons enter the cell via a number of routes: diffusion, active transport by specific transporters, or by incorporation into membranes. As shown in Figure 5.3(c) they are subsequently

Figure 5.3 (a) Examples of hydrocarbons found in petroleum; examples of a (b) methane and (c) cyclic and straight-chain hydrocarbons metabolism pathway.

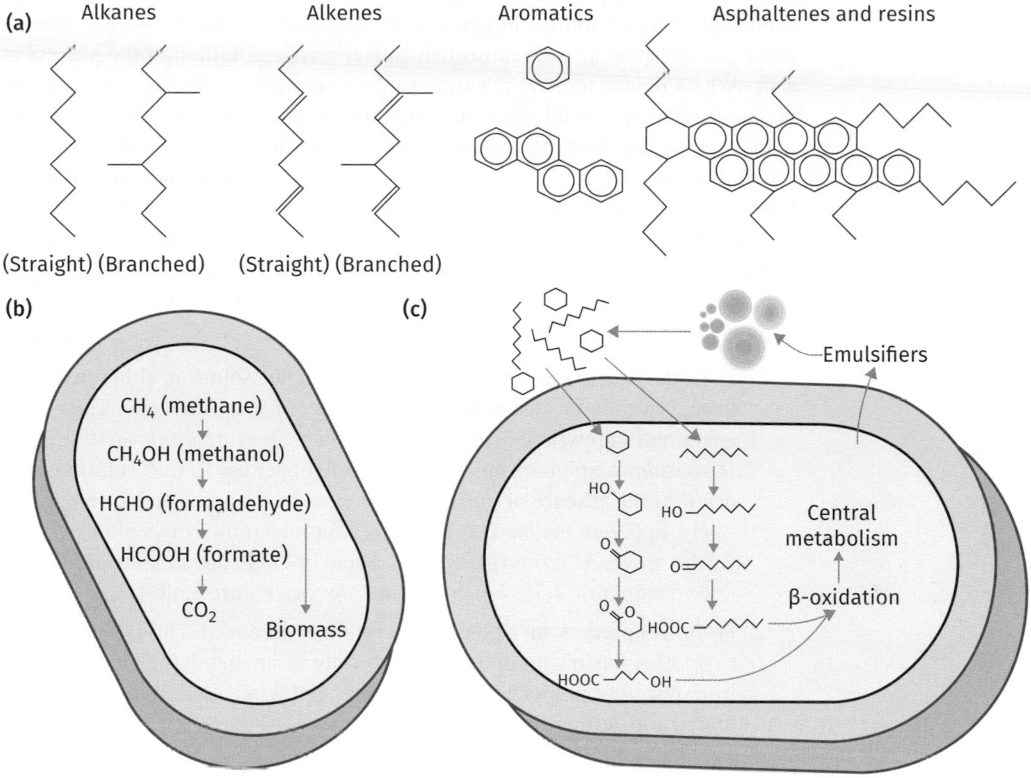

converted to alcohols, then aldehydes, then fatty acids by a series of enzymes. Fatty acids are directed into central metabolism via the β oxidation pathway.

When petroleum spills occur, native populations of hydrocarbonoclastic microorganisms expand to utilize the new energy source. Smaller, less complex hydrocarbons like alkanes and alkenes are metabolized first while larger, more complex resins and asphaltenes take longer to degrade, often lingering in the environment years after the event. During the Deepwater Horizon incident, fertilizer was added to the ocean to spur growth of hydrocarbonoclastic bacteria. Dispersants, predominantly Corexit 9500, were also added in vast amounts (>8 million litres) to break apart the oil. In theory, this should make hydrocarbons more accessible for degradation by individual microorganisms, although there is evidence that Corexit 9500 impedes activity of certain hydrocarbonoclastic bacteria and may itself be toxic.

Key points

- Crude oil is composed of a vast range of different hydrocarbons.
- Native populations of microbes can play an important role in degrading hydrocarbons.
- Adding fertilizer and possibly oil dispersants can spur growth of hydrocarbon-degrading microorganisms.

5.4 Recovery of metals and bioleaching

Bioleaching is the process whereby microorganisms are used to extract metals from their mineral source. In general, the goal is to enrich low-grade ores in order to lower the costs of subsequent metal recovery, although the process is also used to reduce mineral release into the environment. Bioleaching has numerous advantages compared with more traditional extraction techniques using high-temperature roasting and smelting. Under the correct conditions, bioleaching is cheaper, requires less energy, and is more environmentally friendly. However, it is a much slower process and therefore is generally only applied to low-grade ores containing metals of sufficiently high value. There are hundreds of different processes utilized but one is outlined in Figure 5.4 as an example. This shows copper extraction from covellite (CuS), which is currently applied at various mining operations around the world.

1. First, the low-grade ore is treated with an acidic solution, either in a heap, typically located within a large basin, or in a bioreactor. This encourages growth of acidophilic bacteria such as *Acidithiobacillus ferrooxidans*. Spontaneous conversion of copper ore to Cu^{2+} solution can occur in the absence or presence of oxygen (reactions 1 and 2, respectively, in Figure 5.4(a)). Fe^{3+} is required for anaerobic conversion. At this stage, *A. ferrooxidans* plays a role in oxidizing sulphur in CuS to sulphuric acid, which aids the process (Figure 5.4(b)).
2. From this process, an acidic Cu^{2+} solution is produced. This is removed to another basin containing iron, typically scrap metal. Cu^{2+} reacts with iron, resulting in precipitation of copper metal (reaction 3), which is separated and further refined.
3. The remaining acidic Fe^{2+} solution is channelled into another pool where *A. ferrooxidans* oxidizes Fe^{2+} to Fe^{3+} solution (reaction 4), which is subsequently recycled with added sulphuric acid to treat another batch

Figure 5.4 (a) Schematic of the bioleaching process of CuS detailing the reactions mediated by *Acidithiobacillus ferrooxidans*; (b) the external reactions mediated between the microorganism and mineral and (c) the process by which this generates energy in the cell.

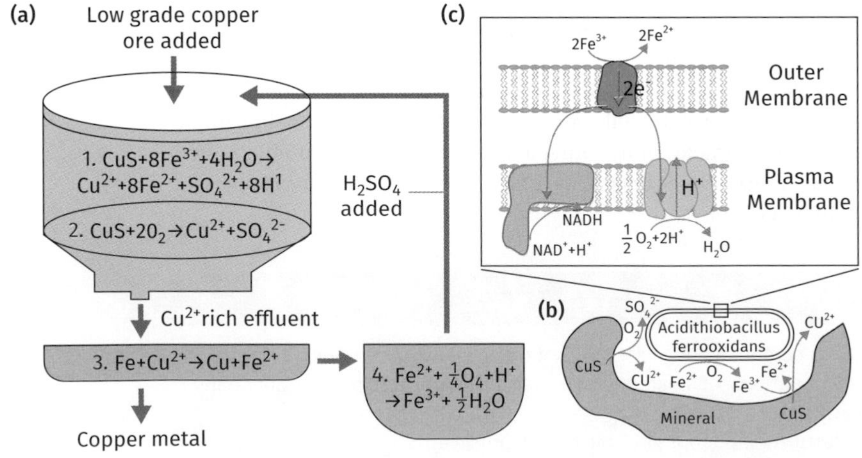

of raw ore. In the process the organism generates electrons which are channelled to acceptor proteins for production of NADH or ATP (illustrated in Figure 5.4(c)). This process occurs naturally in the absence of added acid, albeit at a slower rate, leading to acid mine damage (see Case study 5.1).

Similar processes can be applied to higher-value metals like uranium or gold. Gold is purified from ore using a process called cyanidation, which is detailed in the following equation: $4Au+8NaCN+O_2+2H_2O \rightarrow 4Na[Au(CN)_2]+4NaOH$. However, gold found in arsenopyrite ores (FeAsS) is not readily accessible to cyanide. Therefore acidophiles like *A. ferrooxidans* are added to solubilize arsenopyrite ($2FeAsS[Au]+7O_2+2H_2O+H_2SO_4 \rightarrow Fe_2(SO_4)_3+2H_3AsO_4+Au$), in order to release gold for subsequent cyanidation. A similar process can be used to solubilize certain uranium ores. Ores containing insoluble tetravalent uranium are treated with a solution of $Fe_2(SO_4)_3$, resulting in production of soluble hexavalent uranium, which is readily purified by other methods (e.g. anion exchange). The process is outlined in the following equation: ($UO_2+Fe_2(SO_4)_3 \rightarrow UO_2SO_4+2FeSO_4$). Similar to what occurs in covellite bioleaching, *A. ferrooxidans* plays a role by reoxidizing Fe^{2+} ($FeSO_4$) to Fe^{3+} ($Fe_2(SO_4)_3$).

Case study 5.1
Acid mine damage

Acid mine damage has been a phenomena observed since ancient Roman times. It is caused by exposing sulphide ores to environmental conditions. This results in production of ferrous ions and sulphuric acid which acidify and stain local water catchments (as shown in the Case study Figure CS 5.1 (a)), leading to death and damage of wildlife and plants. Depending on the size of the mining operation, acid mine damage can continue for decades.

How can bioleaching be used to limit acid mine damage?

Figure CS 5.1 (a) A river (Rio Tinto, Spain) stained red by acid mine damage; (b) an example of one of the chemical reactions leading to production of sulphuric acid, in this case pyrite (FeS_2) oxidation.

(a)

(b)

$$2FeS_2+7O_2+2H_2O \rightarrow 2Fe^{2+}+4SO_4^{2-}+4H^+$$

(a) Ronald Karpilo / Alamy Stock Photo

Key points

- Bioleaching is generally used to extract high-value metals from low-grade ores.
- Microbes play a role by regenerating Fe^{3+} solution, generating sulphuric acid or by solubilizing metals.
- Bioleaching can be used to minimize contamination of local environments caused by acid mine damage.

5.5 Environmental monitoring

Environmental monitoring involves using specialized devices to detect and measure compounds, including contaminants or pollutants, in the environment or in a biological sample. Live microorganisms can be incorporated into some of these devices, termed biosensors. **Biosensors** consist of a biological recognition element, integrated with a physical transducer, which generates a signal that allows analysis of the concentration of a specific compound. Biosensors have existed since the 1960s but have predominantly used enzymes, many of them originating from microorganisms or produced via heterologous protein expression in microbes (see Chapter 1, section 1.5), as biological recognition elements. An early example is the glucose biosensor. If you examine Figure 5.5(a) you will observe the basic elements of this device. Glucose, generally from a blood sample, is detected by an enzyme, glucose oxidase, which converts it to gluconic acid. In the process, an electron is released, which is directly transferred to an electrode (transducer). The scale of electron production correlates to the concentration of glucose in the sample, which is indicated in the signal processing system.

Figure 5.5 Schematic detailing how an (a) enzyme based glucose biosensor and (b) microbial based biochemical oxygen demand biosensor functions.

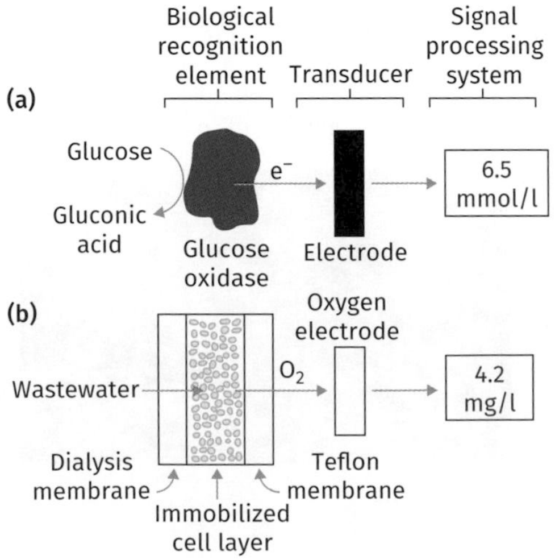

You can learn more about biosensors and how they are used in Chapter 7.

In theory, using microbes instead of enzymes as biological recognition elements offers some advantages. Purification of enzymes are not required and certain species can be easily cultured to high volume. However, sensitivity, selectivity, and response time are major issues, as well as the maintenance and viability of the microbes used. Specifically, these issues must be addressed in order to develop a functional device:

1. Sensitivity: In order to detect the specific compound, either the biological recognition element (typically a protein) must be located at the cell surface or the compound can diffuse or be actively transported into the cell interior. One possible solution is to genetically engineer the organism to display an intracellular protein on the cell surface by attaching it to an outer membrane protein.
2. Selectivity: In the case of the glucose sensor, the electrical signal was derived from the activity of a single enzyme. Microorganisms typically contain hundreds to thousands of active enzymes. Therefore, ensuring that the signal is derived from the correct biological recognition element(s) is more challenging. Again, genetic engineering of an organism can potentially solve this issue. For example, by linking activity of the biological recognition element to a **reporter gene**, such as luciferase, the chemiluminescent signal can be easily differentiated from other cellular activity.
3. Response time: The response from the biological recognition element to the transducer must occur within a sufficient time frame for the data to be useful. In the case of the glucose sensor, the enzyme, transducer, and the signal detector are in close proximity, resulting in rapid analysis. In microbial biosensors, the signal has to be relayed from the cell to the transducer over longer distances. **Immobilization** of microorganisms to the transducer, either by chemical cross-linking or by physical attachment can optimize the response between the two elements.

Due to these difficulties, commercialization of microbial biosensors is not as advanced as enzyme-based systems. An early example, which is outlined in Figure 5.5(b), is the biochemical oxygen demand (BOD) biosensor. This device measures the amount of dissolved oxygen in water required to oxidize all biodegradable organic compounds in a specific volume. A sample, for example from wastewater, is injected into the system. Dissolved organic compounds from this sample diffuse through a dialysis membrane to a layer of immobilized bacteria that metabolize the material, resulting in an increase in respiration and oxygen consumption. The remaining oxygen diffuses through a Teflon layer where it is detected by an oxygen electrode. The difference in oxygen demand before and after the addition of the sample results in a signal which is proportional to the easily biodegradable organic substrate present. However, it should be noted that microbial BOD biosensors have been replaced by models utilizing electrochemical technology. Very few microbial biosensors have been commercialized, with most still at the developmental stage.

Key points

- Microbial biosensors use immobilized microorganisms to detect an external compound.
- Most microbial biosensors are in development with very few being commercialized compared with enzyme-based systems due to sensitivity, selectivity, and response time issues.

5.6 Gas bioscrubbing

Using microbes for gas bioscrubbing is an established industrial technology. Gas bioscrubbing involves removing biodegradable compounds from contaminated air, allowing the cleaned gas to be recycled or released into the environment. The process is currently used to reduce odorous compounds generated by various industries producing, for example, rubber, polymers, enzymes, and cigarettes. It is also used to purify gas streams generated from water purification plants and waste sites. The process, outlined in Figure 5.6, first involves passing polluted air through a scrubber, where contaminants are absorbed from the gas stream. This can be performed by either:

1. Spraying a liquid stream containing a suspension of microorganisms which is subsequently cycled between the scrubber and a bioreactor.
2. Passing the gas stream through wash water where water soluble contaminants are absorbed before being transferred to a bioreactor.

In the bioreactor, a consortium of microorganisms metabolize the contaminants, which can include alcohols (e.g. methanol, ethanol), aldehydes (e.g. formaldehyde), ketones (e.g. acetone), carboxyl acids (e.g. EDTA), phenols, amines, H_2S, and hydrocarbons. Appropriate nutrients and aeration are required. The efficiency of the process is dependent on the ability of the consortia to degrade pollutants and the incubation time. As is the case with anaerobic digestate tanks, introducing new waste streams may require a different consortium of microorganisms for efficient degradation, reducing efficiency. Therefore,

Figure 5.6 Schematic of a gas bioscrubbing facility.

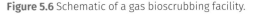

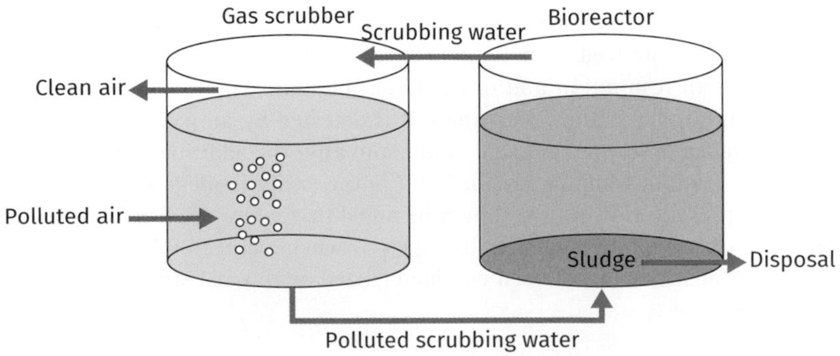

maintaining a consistent waste stream is important. There are a few disadvantages of the system. If pollutants are not water soluble then degradation will be limited. Sludge, predominantly biomass, is also produced in the bioreactor and must be disposed of appropriately.

Key point

- Gas bioscrubbing is an established technology which uses a consortium of microorganisms to remove biodegradable contaminants from waste streams.

5.7 Bioaugmentation

Bioaugmentation is the process of adding foreign microorganisms, which could be genetically modified, to a specific environment to facilitate bioremediation. Typically, cultures are prepared and stored in laboratories and then used when the native microbial population has been depleted due to environmental factors or invasive species, or when it is incapable of degrading certain compounds contaminating a site. Bioaugmentation is commonly used to regenerate populations in wastewater, gas bioscrubbing bioreactors, and in anaerobic digestate tanks. Multiple factors must be considered for success:

1. Will the introduced microbial population survive the environmental conditions?
2. Will the introduced microbes successfully compete with the native microbial population?
3. What is the aliquot size required for successful establishment of a new/modified microbial population?
4. If a genetically engineered organism is introduced into a new environment, appropriate regulatory approval must be obtained.

Prior to bioaugmentation, some knowledge of the environmental conditions is required, for example: pH, temperature, contaminant variation and concentrations, nutrient levels, and existing microbial populations. Suitable microbes are stored as powders or freeze dried pellets. These are typically generated from laboratory cultures prepared by enriching organisms grown in media with the contaminant of interest or pure cultures consisting of an organism adapted or genetically modified to degrade a specific compound. Appropriate microorganisms are introduced into the selected environment either as high-density cultures or encapsulated in carriers such as agarose or microbeads. Successful adaption to the new environment is not guaranteed and monitoring is required in order to assess the impact of the introduced microorganisms. A common example is bioremediation of wastewater plants. Many companies now offer products which include a mixture of microorganisms, enzymes, and nutrient supplements (e.g. Suez BioPlus), designed to regenerate or enhance native populations, along with the expertise to monitor the results following treatment. Other companies offer similar products for bioremediation of sites polluted with petroleum and polycyclic aromatic hydrocarbons (e.g. MicroGen Biotech, Dunton Environmental Ltd).

Key points

- Bioaugmentation is typically used to regenerate or enhance microbial populations in wastewater treatment, anaerobic digesters, or gas bioscrubbing bioreactors.
- Bioaugmentation is used for bioremediation of sites polluted with hydrocarbons.

Chapter Summary

- Wastewater treatment is an established and commercially viable technology used to treat water from a range of commercial, household, and agricultural sources.
- Wastewater treatment is a multi-stage process. Microbes play an active role in the aerobic oxidation of organic matter and treatment of sludge.
- Anaerobic digestion is an established and commercially viable technology used to convert various waste streams to methane and digestate.
- Anaerobic digestion involves four processes: hydrolysis, fermentation, acetogenesis, and methanogenesis.
- Hydrocarbon-degrading microorganisms degrade certain petroleum compounds released into the environment, which can be enhanced by adding fertilizers.
- Bioleaching is a commercially viable technology used to extract metals from low grade ores.
- Acidophilic bacteria aid bioleaching by producing sulphuric acid and regenerating Fe^{3+}.
- Microbial biosensors are still at the developmental stage.
- Gas bioscrubbing is an established and commercially viable technology for removing biodegradable contaminants from waste streams.
- Gas bioscrubbing uses a consortium of microorganisms in a bioreactor to remove contaminants.
- Bioaugmentation is typically used to replace depleted microbial populations in wastewater and gas bioscrubbing bioreactors and in anaerobic digestate tanks.
- Bioaugmentation can involve introducing a foreign microbial population into an environment to degrade specific contaminants, i.e. hydrocarbons.
- Using genetically modified organisms for bioaugmentation will require regulatory approval, which may be challenging in many countries.

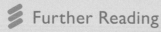

Further Reading

Bitton, G. (2011). *Wastewater Microbiology*, 4th edn.

This book provides detailed explanations and examples of wastewater remediation.

Tyagi, M., da Fonseca, M. M. R., and de Carvalho, C. C. C. R. (2011). 'Bioaugmentation and biostimulation strategies to improve the effectiveness of bioremediation processes'. Biodegradation 22: 231–41.

This paper provides more details on various strategies of bioremediation.

Venkiteshwaran K., et al. (2015). 'Relating anaerobic digestion microbial community and process function'. Microbiol. Insights 8: 37–44.

This paper gives a more detailed insight into the process of anaerobic digestion.

Discussion Questions

5.1 What are possible low-cost solutions for wastewater treatment in rapidly urbanizing cities in developing countries?

5.2 What future research is required in order to understand the best approach to clean up oil spills?

5.3 Are genetically modified organisms a realistic approach to improve processes such as wastewater treatment, anaerobic digestion or hydrocarbon remediation?

6 APPLICATION OF SYNTHETIC BIOLOGY TO BIOTECHNOLOGY

Learning Objectives

- To define synthetic biology;
- to identify the main organisms that are used and why;
- to understand the importance and issues of modelling metabolism and fluxes;
- to understand the various methods by which organisms can be genetically manipulated;
- to describe mechanisms to improve production of compounds or proteins;
- to list potential applications of synthetic biology;
- to detail the issues and the scale of production required for a range of different valued products.

Synthetic biology is a rapidly growing field of research which attempts to apply engineering principles to biological organisms. There are a number of definitions which can be used to describe synthetic biology. The UK Royal Society defines it as 'the design and construction of novel artificial biological pathways, organisms or devices, or the redesign of existing natural biological systems.' A report commissioned by the European Union defined it as 'the engineering of biology: the synthesis of complex, biologically based (or inspired) systems which display functions that do not exist in nature.' Regardless of the exact definition, synthetic biology is a technology that has the potential to transform microbial biotechnology. It may allow development of organisms that can produce a wide variety of useful compounds, ranging from biofuels to industrial chemicals, therapeutic medicines (vaccines, drugs, antibodies, biologics), biosensors and biomaterials, in addition to microbial strains with improved bioremediation, agricultural, and medical properties.

This chapter will explore a range of topics in this area, including the main microorganisms used, the methods by which species are 'designed' to fit the desired function, production processes, and the range of products that are and could potentially be produced via synthetic biology approaches. Since synthetic biology is an applied science and the desired outcome is often the production of a commercially viable product, economic and industrial factors are considered. Issues with the technology, and successes and failures in the field will also be discussed.

6.1 Microorganisms widely used in synthetic biology

Synthetic biology may be applicable in a range of organisms and cells, but this chapter will focus primarily on bacteria, fungi, and microalgae. In order to 'engineer' microorganisms there are two major requirements. First, a certain degree of knowledge about the organism's biology, particularly metabolism, gene regulatory networks, and transport of compounds in and out of cells, is required. Second, it is necessary to be able to genetically manipulate an organism, preferably on a repeatable basis, in order to modify it to perform the desired function, e.g. production of biofuels. These requirements massively reduce the number of suitable species. *Escherichia coli* and *Saccharomyces cerevisiae* are still the organisms of choice for the majority of synthetic biology applications due to a number of advantages. Genetic manipulation is relatively simple and it is easy and quick to grow cultures for both research and industrial applications. Other organisms of choice include the soil bacterium *Corynebacterium glutamicum*, an industrial strain used to produce amino acids such as glutamate and lysine, and bacterial *Bacillus* and the filamentous *Aspergillus* species, for production of industrial enzymes found predominantly in laundry detergent, food, and pharmaceuticals. All of these species require an organic carbon source, typically sugars, for production of these compounds. However, photoautotrophic microorganisms such as the cyanobacterium *Synechocystis* sp. PCC 6803, or the green alga *Chlamydomonas reinhardtii*, can convert carbon dioxide to a range of useful products using energy derived from photosynthesis, which may offer advantages for long-term, stable production of desired products without the requirement for an agriculturally derived carbon source.

Key points

- For synthetic biology applications, organisms which can be repeatedly genetically manipulated are optimal.
- Ideally, organisms which are highly characterized should be chosen for synthetic biology.

6.2 Metabolic pathway modelling

In order to manipulate an organism in the same way that an engineer would modify a man-made device, a knowledge of the various parts and processes is necessary. In the case of a microorganism, this requires knowledge of the biological processes in the cell, particularly metabolic **fluxes**, transport pathways, and regulation of gene networks. The main issue is that while an engineer typically understands all the components in a device and how these interact with each other, our knowledge of even the most widely studied and best understood organisms, such as *E. coli* and *S. cerevisiae*, is incomplete. For example, in *E. coli* strain K-12, the primary organism used for laboratory studies, approximately 20 percent of genes are of unknown function. Our knowledge of how various cellular metabolites partition between different biosynthetic pathways and the regulation of these processes is even poorer. Therefore, mathematical modelling of metabolism, a process termed **flux balance analysis**, is applied. This relies on building an interactable information network consisting of all the known processes within the cell, a time-consuming process since it depends on reviewing the entire scientific literature (typically thousands of papers). More specifically, the network would need to contain information on metabolic reaction stoichiometry, chemical formulae and charges of metabolites, the association between genes, proteins, and reactions, and the location of components within different compartments of the cell. An example of the core *E. coli* metabolic network is shown in Figure 6.1. Note that many of the metabolic pathways are heavily interconnected. This also only represents a small proportion of the entire network and has no information on gene regulation, which demonstrates the complexity involved in modelling a biological organism. The great advantage of this system is that once established, investigating the fluxes is extraordinarily rapid compared to more laborious laboratory experiments for tracking metabolism.

Once a metabolic database is compiled, the next step is to mathematically represent metabolic reactions by tabulating the stoichiometric co-efficiencies of each reaction. This involves imposing constraints in terms of inputs (i.e. substrate, ATP) and outputs (i.e. products) for each reaction, which will affect other parts of the network. The key aspect is to ensure that production of specific compounds is balanced by consumption. A computer is then used to calculate the optimal solution to a given function, usually derived from data generated from growth experiments. This is constrained by the known metabolic network and a number of user-generated inputs or assumptions (i.e. carbon source or

Figure 6.1 Map of the core *E. coli* metabolic network. Detailed are a small proportion of the reactions in this species. The blue arrows indicate reactions, orange circles—cytosolic metabolites, yellow circles—extracellular metabolites, and the enzymes are in upper case.

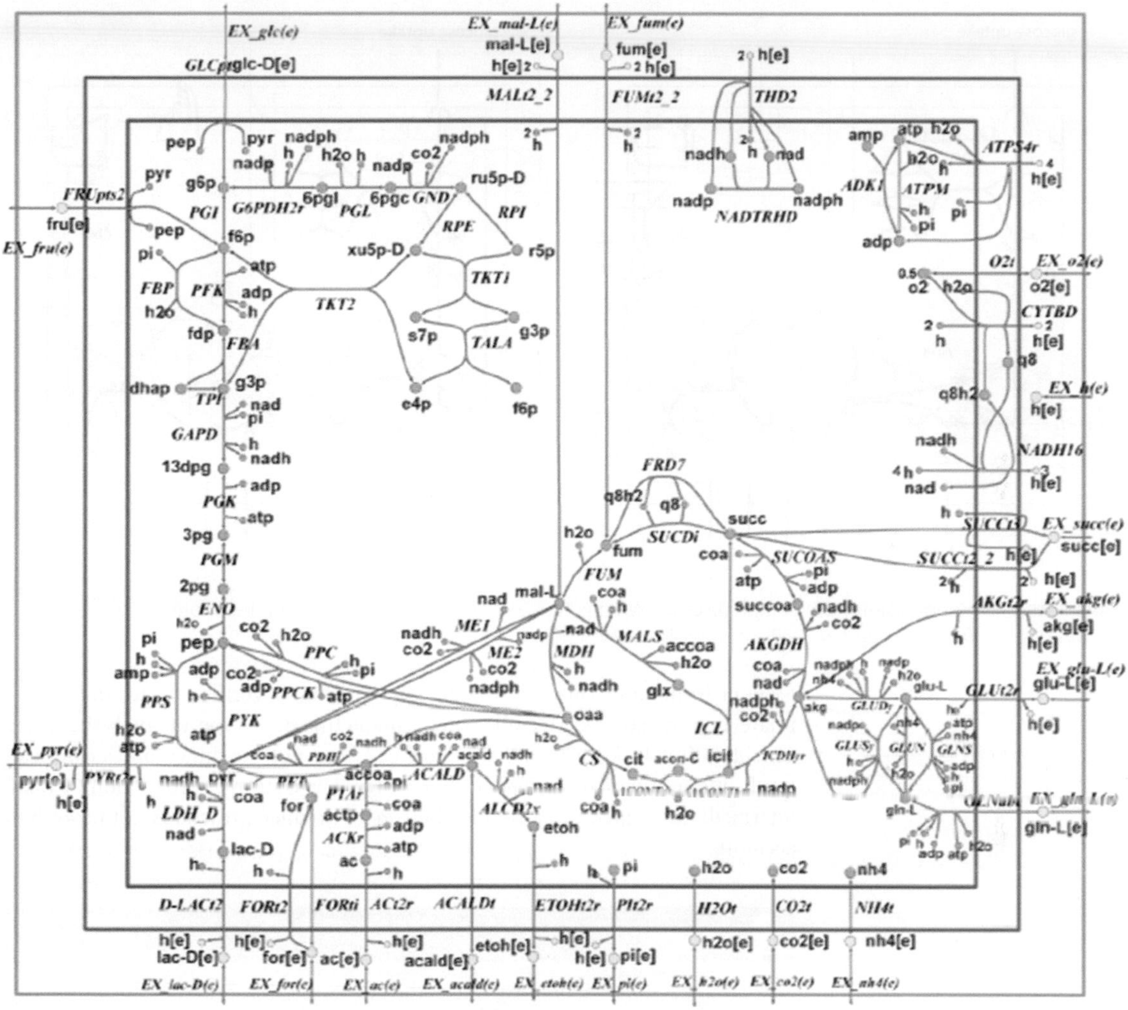

availability of oxygen). For example in Figure 6.2 the metabolic flux of *E. coli* cells cultured under aerobic and anaerobic conditions is shown. Notice that while the cell uses some metabolic pathways for both processes, including metabolism of glucose, others are unique to one or the other.

Flux balance analysis can also be applied to improve a specific outcome, usually production of a desired metabolite. For example, if the substrate is glucose and the desired product is lactate, which is derived from anaerobic fermentation, flux balance analysis can be used to determine the amount of lactate expected from a certain input of glucose. It also allows the user to predict whether altering other metabolic reactions (i.e. by adding or deleting genes) will modify

Figure 6.2 Metabolic flux under (a) aerobic and (b) anaerobic conditions. The map is the same as the one in Figure 6.1. Highlighted in dark blue are the active reactions under each growth condition. PPP—pentose phosphate pathway; Glyc—glycolysis; Ana—anaplerotic reactions; TCA—tricarboxylic acid cycle; Ferm—fermentation pathways; OxP—oxidative phosphorylation.

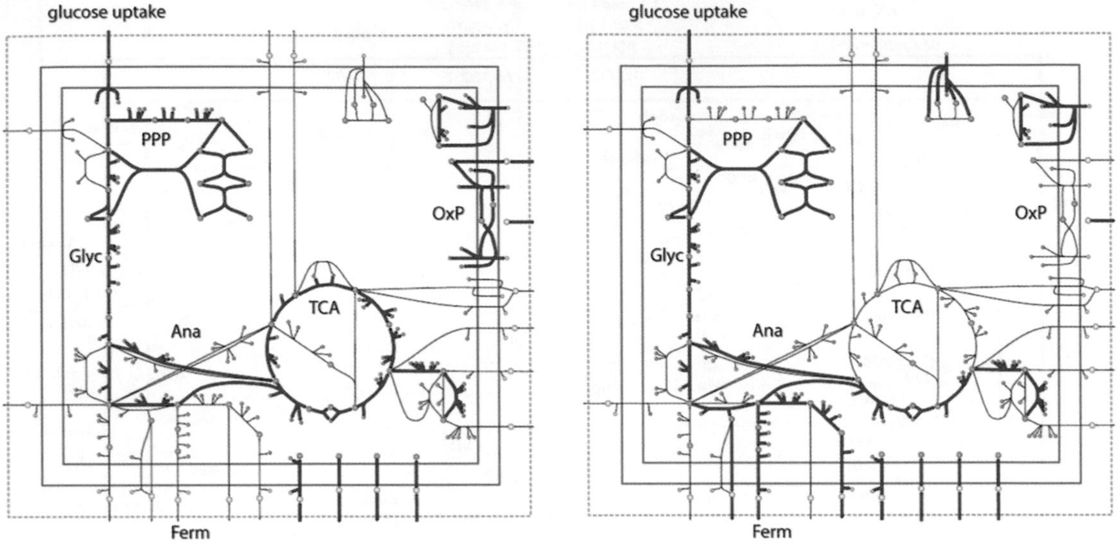

product formation. This can then be experimentally verified by testing appropriate mutant strains. Overall, when conducting this type of analysis, a key consideration to keep in mind is that cells are not machines and errors can result due to a lack of biological knowledge. For example, production of toxic intermediates, which can alter metabolism and reduce growth, is not taken into account.

Key points

- An understanding of metabolic pathways, fluxes, and cellular transport and regulation is advantageous in order to genetically modify an organism.
- Flux balance analysis can be used to rapidly model the effect of altering various conditions in the organism or environment.
- Modelling must be verified via experimental methods.

6.3 Gene synthesis

Another key development underlying synthetic biology is the rapid development and decreasing cost of chemically synthesizing fragments of DNA. Current methods rely on chemical synthesis of short oligonucleotides, which are then assembled into longer DNA strands, up to ~10 kb, via PCR, ligation, or recombination. The key advantage of gene synthesis is the ability to generate a custom strand of DNA, without having to rely on sequences found in nature, which is required for PCR. Prices can be as low as US$0.07 per base, although repeat sequences and those with high GC content cost more. In some cases, it is now cheaper to synthesize a plasmid or DNA sequence than to generate it using classical cloning techniques (i.e. PCR, digestion, ligation, transformation of cells).

Key point

- Gene synthesis is a relatively inexpensive method of constructing fragments of DNA.

6.4 Using biological parts to facilitate genetic manipulation of microbes

In order to engineer appropriate mutants, precise genetic manipulation is required. Basic principles involved in genetic manipulation, including design of plasmids, inserting plasmids into cells, chromosomal integration, and gene expression are outlined in Chapter 1, section 1.4. In the case of synthetic biology, a key goal is to design interchangeable parts which make it easier and quicker to build multi-component genetic systems, generally plasmids, in the organism of choice. An example is shown in Figure 6.3.

In the case shown in Figure 6.3, the components, sometimes called biobricks, consist of a range of promoters (P), genes (G), and terminators (T). Each biobrick contains a standard part, called a prefix and suffix, flanking the region of interest which facilitate assembly— much like joining together pieces of Lego. The prefix and suffix incorporate restriction endonuclease sites or overlapping sequences that allow for recombination. Biobricks can be synthesized, amplified

Figure 6.3 Use of biobricks to assemble a plasmid library.

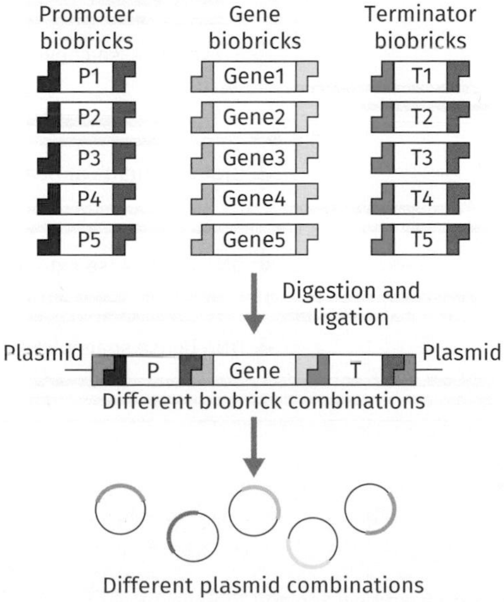

using PCR, or excised from pre-existing plasmids found in repositories such as the registry of standard biological parts (http://parts.igem.org/Help:2017_DNA_Distribution) (see Case study 6.1). Gene cassettes with the appropriate restriction endonuclease sites can then be assembled together in the correct combination (plasmid-gene-terminator) and inserted into appropriate plasmids by standard digestion and ligation techniques. In the case of recombination, a different technique is used called Gibson assembly. The steps are outlined in Figure 6.4. Different fragments containing overlapping ends are treated with an **exonuclease**, which digests the 5' ends of dsDNA. The homologous regions of the fragments anneal, after which DNA polymerases repair any gaps. The fragments are then ligated together. Multiple fragments can be assembled together into a new plasmid rapidly within a single tube. The plasmids can then be transformed into the organism of interest and the library screened for the desired phenotype, e.g. increased metabolite production. More biobricks, i.e. **enhancers**, flanking regions for chromosomal integration, can be incorporated into plasmids when required.

Automated systems can now perform most of the functions required for assembling a plasmid library. This includes PCR, fragment assembly, transformation,

Figure 6.4 Gibson assembly of DNA fragments. Fragments of DNA with homologous regions are generated by PCR and incubated together at 50°C. Three enzymatic activities: exonuclease chewing back of 5' ends leading to subsequent annealing of homologous regions; DNA polymerase filling of gaps; and ligation of annealed fragments, is required for joining of fragments into the final plasmid.

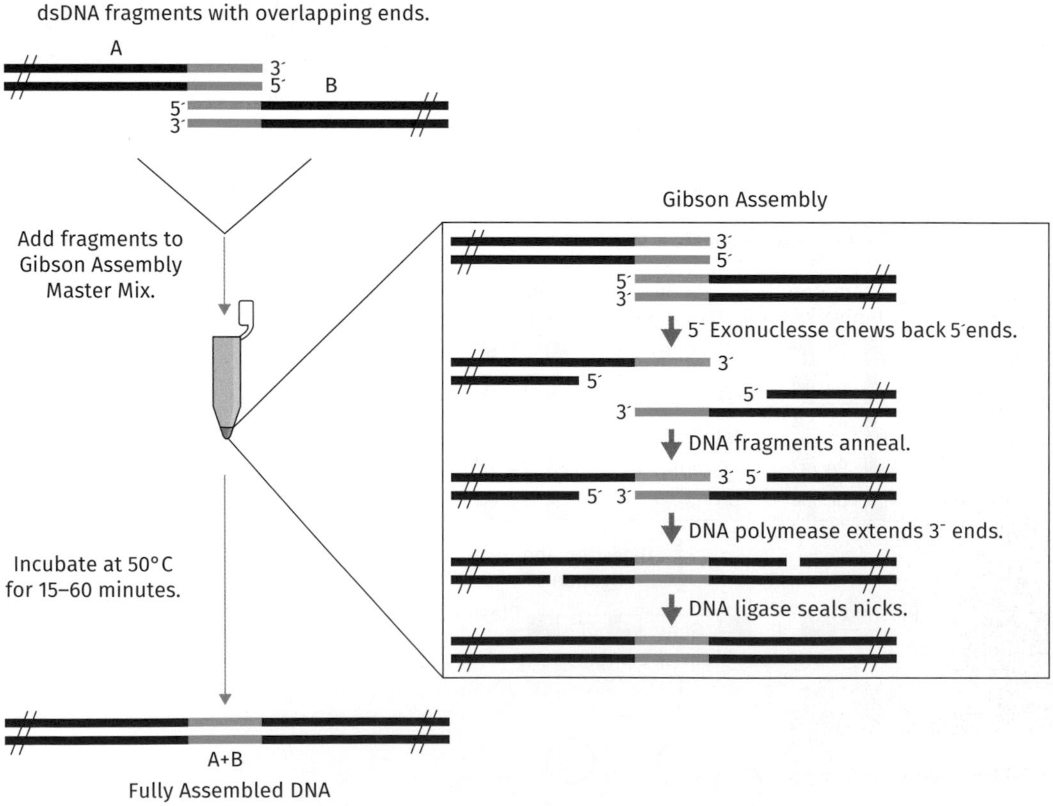

plasmid purification, and sequencing to confirm whether the plasmid is correct. It is now possible to generate libraries containing thousands of plasmids rapidly using such systems.

Key points

- Biobricks can be used to rapidly assemble DNA fragments together to form gene cassettes.
- Gibson cloning can be used to assemble overlapping DNA fragments together via homologous recombination.

Case study 6.1
The iGEM competition

In addition to housing a registry of biological parts, iGEM conducts an annual, worldwide, synthetic biology competition aimed at undergraduate university students at selected institutions. Multidisciplinary teams work on a wide range of projects related to synthetic biology, for example systems for controlling population of different microbes in a consortia, or development of a cheap, user-friendly biosensor that can detect poisons in medical products. Recent projects are highlighted on their website: http://igem.org/Competition, and are worth investigating if students are at an institution that participates in this competition and would like to be involved.

If you are interested in this competition think about possible projects that would be suitable for the iGEM competition based on past successful projects.

What disciplines would be necessary to develop a successful iGEM team?

Figure CS 6.1 Image of students attending the 2016 iGEM competition.

iGEM Foundation and Justin Knight

6.5 Directed evolution of cells

An alternative approach to modifying organisms is directed evolution, whereby cells are mutagenized in order to develop strains better suited to certain environmental stresses or which produce higher amounts of a desired product. For example, above a certain concentration ethanol is toxic to most microorganisms. Ethanol-tolerant strains have been generated by culturing cells (mostly yeast), which are randomly mutagenized by either chemical treatment or ultraviolet light, in media containing gradually increasing concentrations of ethanol. Cells that develop mutations allowing them to adapt to this environment will survive and dominate the culture after successive generations.

In contrast with classical mutagenesis of microbial strains, modern synthetic biology approaches seek to direct evolution of organisms via targeting specific genes. Multiplex Automated Genome Engineering (MAGE) is one such example, with the different steps outlined in Figure 6.5. First, a synthetic oligonucleotide

Figure 6.5 Multiplex Automated Genome Engineering.

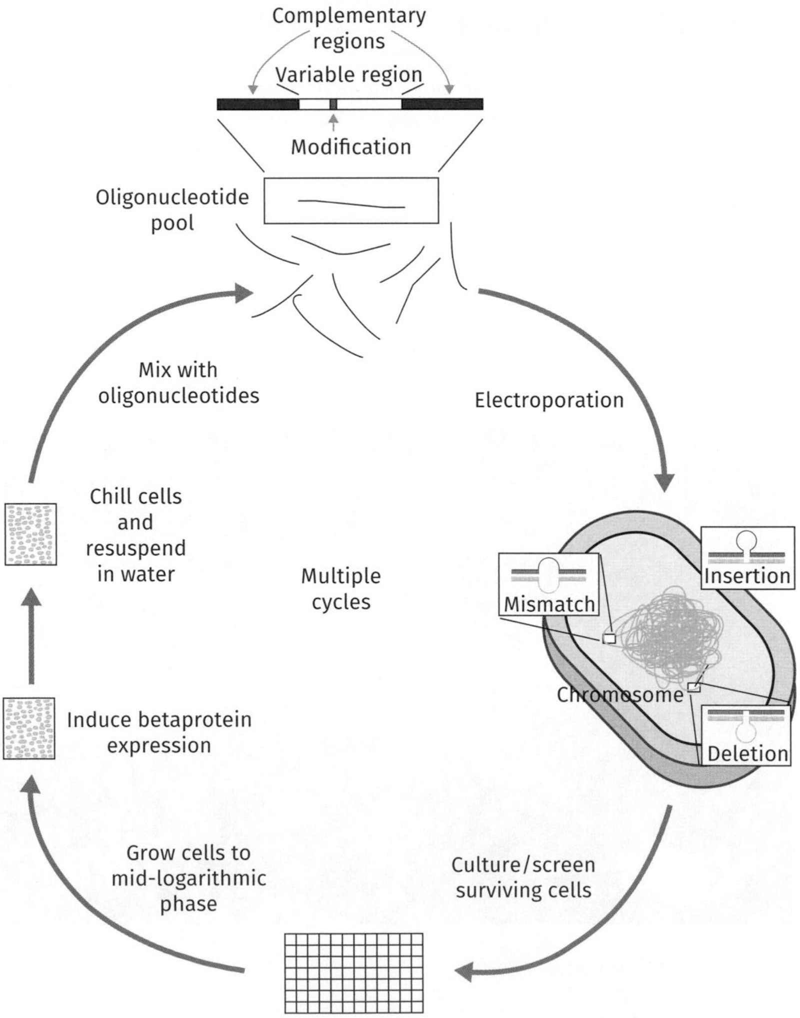

library is constructed incorporating a sequence that will introduce the desired mutation(s) (i.e. nucleotide insertion, deletion, or modification), flanked by sequences complementary to the genome region in which the modification is to be introduced. Cells are first grown to a sufficient population. They are then incubated at 42° C to induce expression of a specific beta protein which facilitates oligonucleotide binding to chromosomal DNA, and subsequently at 4° C to minimize toxic side effects. The oligonucleotides are then introduced into the cell via electroporation. Once inside the cell, they hybridize to complementary sequences, inducing the required mutation in a percentage of cells. Multiple oligonucleotides can be introduced simultaneously to target different regions of the chromosome. The process is then repeated for a certain number of generations, resulting in a population of cells with different genetic variation at each point of the chromosome targeted. Due to the vast number of variants, a method for screening or selecting colonies of interest is required. MAGE has been successfully applied to certain *E. coli* strains expressing the beta protein and which lack **mismatch repair**. However, its use in other microorganisms has been limited since the process has to be optimized for each species.

Key points

- Directed evolution can be used to simultaneously mutate different regions of the chromosome.
- Multiplex Automated Genome Engineering is currently limited to a few bacterial species.

6.6 Modular protein assembly

In model species such as *E. coli* and *S. cerevisiae*, deletion of genes is relatively straightforward. However, optimal expression of a metabolic pathway is more challenging. The efficiency of such a process (i.e. amount of compound produced) is in part determined by spatial dynamics (i.e. the location in the cell of enzymes in a metabolic pathway in relation to each other). Ideally, the enzymes catalysing each step should be in close proximity so that the substrate/product from each reaction is rapidly transferred between proteins. This can be an issue when proteins (typically tens of nanometres in diameter) are expressed in the cytosol, which is a vast area in comparison, since microbial cells are generally microns in diameter. The cytosol is also packed with other proteins, DNA, ribosomes, and various storage bodies for compounds such as carbohydrates and lipids, which limit movement. In this section, we will discuss one mechanism for optimizing the arrangement of proteins to increase metabolite production. An alternative mechanism is discussed in the section on the compartmentalization of metabolic processes (section 6.7).

Certain proteins, such as polyketide synthases, are multi-domain enzymes that catalyse biosynthesis of large, complicated compounds, including antibiotics such as erythromycin. The proteins essentially act as assembly lines, and consist of multiple domains, each adding or modifying a part to the molecule to form the final compound. It is now possible to generate artificial assembly lines using a similar principle to polyketide synthases. If you examine Figure 6.6 you will observe an example. Three enzymes: AtoB, HMGS, and HMGR, which convert acetyl-CoA to mevalonate, a precursor of many commercial compounds,

Figure 6.6 Schematic of a synthetic protein scaffold for improving production of mevalonate.

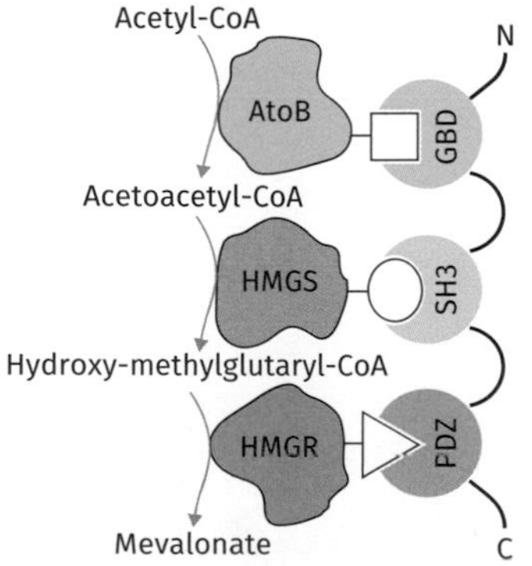

Adapted by permission from Springer Nature: Nature Biotechnology. Synthetic protein scaffolds provide modular control over metabolic flux by John E Dueber et al, 2 Aug, 2009

were covalently attached to a scaffold. The scaffold, encoded on a single open reading frame, contains three protein-protein interaction domains (GBD, SH3, PDZ), which in this case were derived from different mouse proteins. These domains bind specific peptides. Each of the specific peptides was fused to the C-terminus of a different enzyme in the pathway: the GBD peptide to AtoB, the SH3 to HMGS, and the PDZ to HMGR. This results in the binding of the enzymes to the scaffold via the peptide:protein interaction, bringing them into close proximity to each other. An advantage of this system is that the stoichiometry of the three enzymes can be adjusted by altering the number of domains in each scaffold. Therefore, if one enzyme is catalytically slower, a higher number of domains can be added to the scaffold to ensure that more of the corresponding enzyme is attached. In this study adjusting the ratio of AtoB:HMGS:HMGR to 1:2:2 resulted in the highest production, approximately 77-fold higher than non-scaffolded expression.

Key points

- Optimal production is dependent on efficient exchange of metabolites between enzymes in a biosynthetic pathway.
- Enzymes can be co-localized via attachment to a protein scaffold.
- Enzymes attach to the scaffold via peptide:protein associations.
- Protein scaffolds can be used to control the stoichiometry of enzymes in a metabolic pathway.

6.7 Compartmentalization of metabolic processes

An alternative approach to optimizing production of a desired compound in a cell is to localize the relevant proteins and metabolites within a subcellular compartment. Subcellular compartmentalization is a feature of certain bacteria and all eukaryotes, i.e. organelles such as mitochondria. In this section, we will focus on bacterial compartments. These are composed of pentagonal and hexagonal proteins which self-assemble to form a shell. Depending on the properties of the shell, it will be permeable to a limited pool of metabolites. The interior can house proteins or compounds, thus separating them from the bacterial cytosol. For example, the **carboxysome**, the most widely studied compartment, concentrates CO_2 within its interior, where it is incorporated into sugars. If you examine Figure 6.7(a) you will see an electron micrograph of carboxysomes, which are very large compartments (~80–150 nm in diameter) in the cytosol of a cyanobacterium. The schematic details the reactions that occur within this compartment. The carboxysome is permeable to small, negatively charged compounds such as HCO_3^-, 3-phosphoglycerate, and D-ribulose 1,5-bisphosphate. In the carboxysome, HCO_3^- is converted by the enzyme carbonic anhydrase to CO_2, which is incorporated by another enzyme, RuBisCO, into D-ribulose 1,5-bisphosphate to produce 3-phosphoglycerate. This is subsequently released from the compartment and funnelled into glycolysis. RuBisCO can also catalyse a reaction between O_2 and D-ribulose 1,5-bisphosphate, resulting in production of the toxic compound glycolate 2-phosphate, which must be converted to 3-phosphoglycerate by an energetically expensive series of reactions.

Figure 6.7 Manipulating bacterial compartments for synthetic biology. (a) Electron micrograph of a carboxysome and a schematic of the reactions occurring in this compartment; (b) Construction of an artificial compartment.

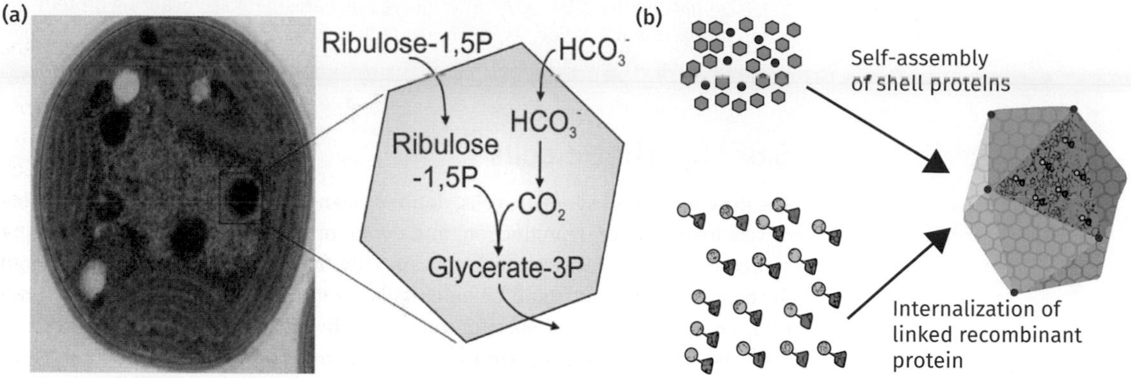

By concentrating CO_2 the carboxysome minimizes glycolate 2-phosphate production, greatly improving the efficiency of carbon fixation.

Potentially, compartments could be used for a range of synthetic biology applications, which are similar to the roles they perform in bacteria. Some compartments house enzymes that are toxic to the cell or produce toxic compounds. For example, the Eut compartment contains enzymes that converts ethanolamine to ethanol, acetyl-phosphate, and acetyl-coenzyme A. An intermediate compound, acetaldehyde, is toxic. By enclosing these catalytic reactions within the compartment, acetaldehyde is not released into the cytosol, thereby mitigating negative effects on the cell. From a synthetic biology perspective, introduced metabolic pathways that also produce toxic intermediates may have fewer deleterious effects on the cell if the enzymes are confined to a compartment. Likewise, expression of proteins that are toxic to cells is a major issue, limiting production. Compartments have been engineered in microbial recombinant protein expressing hosts such as *E. coli* to house foreign toxic proteins, for example, HIV protease, which allows for higher production.

In order to design synthetic compartments, there are a number of requirements, which are outlined in Figure 6.7(b). It is necessary to express all the proteins that form the shell of the compartment. Since the components of the protein shell self-assemble, the major requirement is to ensure sufficient expression of the required genes. Carboxysome shell proteins have been expressed in *E. coli* and structures similar to the assembled compartment have been observed. Conversely, if the bacterium of interest already contains a compartment, this could be used by removing the native internal proteins, thus creating a space in which recombinant proteins could be inserted. Next, the proteins of interest have to be modified so they will assemble within the compartment. Our understanding of how native proteins assemble within compartments is limited. However, certain peptide sequences have been shown to internalize proteins within some compartments. Conversely, recombinant proteins have been fused to internal compartment proteins. Finally, shell proteins may have to be modified to allow for selective transport of desired compounds between the cytosol and compartment interior. This is not yet possible so researchers rely on shell proteins with the desired properties, whilst trying to develop synthetic proteins that may fulfil this role.

Key points

- Compartmentalization is an alternative mechanism for co-localizing proteins.
- Compartmentalization is a more technically challenging method than attaching proteins to a scaffold.

6.8 Synthetic cells

The development of synthetic cells, defined as an artificially designed cell capable of self-maintenance, reproduction, and evolution, is one of the ultimate goals of synthetic biology. These cells can contain only the essential components required for core functions, but could be potentially modified to perform a distinct function, i.e. production of a specific industrial chemical. In theory, synthetic cells could be a more efficient platform for production of a desired compound, since energy and resources are not diverted towards non-essential functions.

The first step in designing a synthetic cell is synthesizing the chromosome containing all the required genes. To date, such efforts have focused on constructing chromosomes based on free living bacterium with the lowest number of essential genes. Here we will detail a recent, high-profile example in which a chromosome was constructed containing 473 essential genes of *Mycoplasma mycoides*. If you examine Figure 6.8 you will observe the steps in this process, which essentially involved combining overlapping pieces of DNA of gradually increasing size until an entire *Mycoplasma* chromosome was assembled. The first step was to synthesize oligonucleotides with overlapping sequences. Approximately 48 oligonucleotides were pooled into different groups and used to generate a series of ~1.4 kb fragments by PCR. These 1.4 kb fragments have overlapping sequences, which facilitates assembly into ~7 kb fragments and insertion into plasmids via Gibson cloning (outlined in section 6.4). Up to 15 7-kb fragments, each with overlapping regions are excised from the plasmids, then transformed into yeast (*Saccharomyces cerevisiae*), where they recombine together via homologous recombination to form a larger DNA fragment, approximately 1/8th the size of the *M. mycoides* genome. These 1/8th fragments, each with overlapping regions, are in turn excised, then recombined together to form the complete *M. mycoides* genome. This genome, which has all the cellular machinery required to function as a replicating chromosome, is then transplanted into another mycoplasma, *M. capricolum*, which had its native chromosome removed. This synthetic cell was able to grow and divide in the same manner as a native *Mycoplasma*.

Synthesis of the whole *Mycoplasma* genome took less than three weeks but cost more than 10 million USD, putting this technology out of the price range of most laboratories. At this stage only a few groups have the skills and resources capable of generating such synthetic microbial strains.

Key points

- Synthetic chromosomes can be assembled from small fragments of synthesized DNA using a combination of Gibson cloning in *E. coli* and recombination in yeast.
- Synthetic chromosomes can be transplanted into certain microbes.
- Generating synthetic cells is an expensive and specialized skill inaccessible to most scientists.

Figure 6.8 Generation of a synthetic *Mycoplasma* cell.

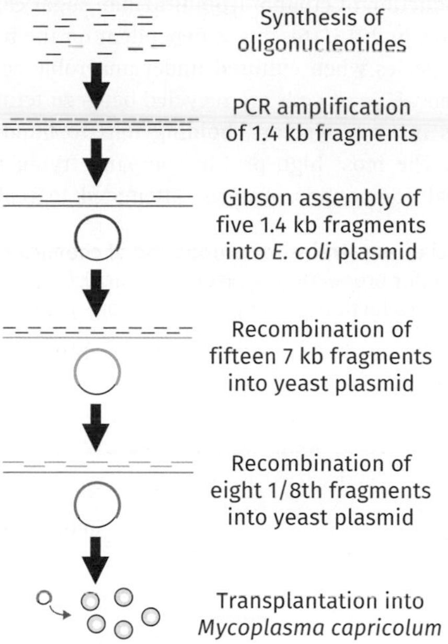

Adapted from Hutchison, C. A., et al., Design and synthesis of a minimal bacterial genome. Science. Mar 25, 2016. Vol. 351, Issue 6280, aad6253. Copyright © 2016, American Association for the Advancement of Science

6.9 Translation of synthetic biology to commercial applications

Engineering organisms is only one aspect which has to be considered when applying synthetic biology to commercial applications. Economic and engineering factors are also important. Some of these (choice of substrate, scale-up, optimization) are discussed in Chapter 1, so here we will focus on the synthetic biology issues using several case studies as examples. The first is the engineering of *Saccharomyces cerevisiae* to produce artemisinic acid, arguably one of the great successes in the field. Artemisinic acid is a precursor for artemisinin, an anti-malarial drug extracted from the plant *Artemisia annua*. If you look at Figure 6.9(a) you will see a metabolic pathway indicating the alterations introduced into the yeast strain. Four genes encoding proteins required for artemisinic acid production (highlighted in green) from farnesyl pyrophosphate (FPP) were expressed in *S. cerevisiae*. In addition, multiple genes encoding proteins in the FPP biosynthetic pathway were up-regulated (highlighted in blue). This resulted in production of 100 mg L^{-1} of artemisinic acid. Further optimization of strains has improved yields to 25 g l^{-1} and production now costs between 350–400 USD per kg, which is undertaken in small reactors (Figure 6.9(b)) containing hundreds to thousands of litres of liquid. Prior to development of this method, artemisinin supplies were erratic and the price varied between $US300–1200 per kg. However, in recent years the cost of plant-derived artemisinin has fallen to less than 250 USD per kg, making the synthetic biology process commercially uncompetitive. None was produced in yeast in 2015.

Applying synthetic biology concepts to develop microbial strains for production of lower value products such as industrial compounds or biofuels is even more challenging. Most of these chemicals are valued at less than $US10 per kg.

While it is possible to produce low-value chemicals using microbes, a good example being production of ethanol from Brazilian sugar cane by *S. cerevisiae* (over 30 billion litres in 2015/16), this is dependent on the native fermentation properties of this species when cultured under anaerobic conditions. To make the process economical, yeast cells are recycled between fermentation runs and the process occurs in large reactors, holding tens to hundreds of thousands of litres of liquid. The most high-profile company trying to synthesize biofuels and industrial compounds, Amyris, attempted to replicate this process

Figure 6.9 Commercializing strains for production of chemicals. (a) Alterations made to S. cerevisiae for production of artemisinic acid; (b) Small bioreactors used for artemisinin production; (c) Amyris bioreactors used for production of farnesene.

(a)

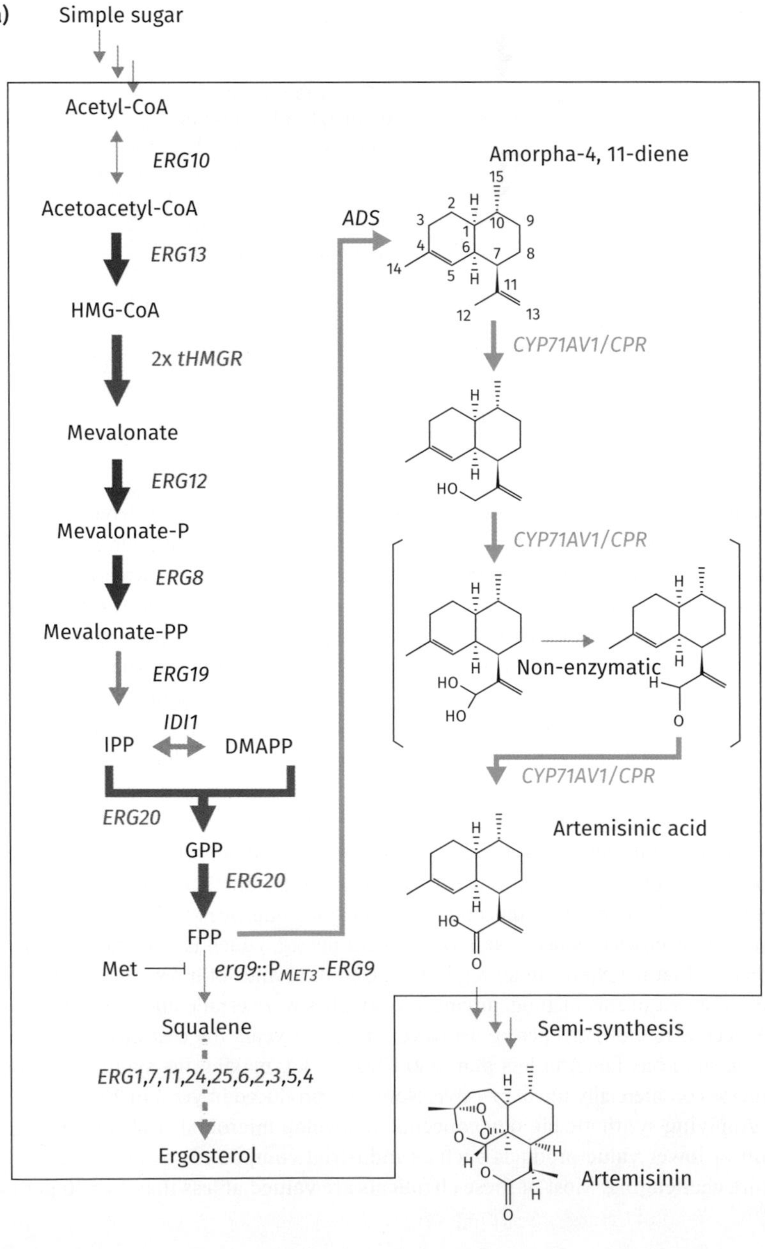

Figure 6.9 (*Continued*)

(b)

(c)

(a) Reprinted by permission from Springer Nature: Nature. Production of the antimalarial drug precursor artemisinic acid in engineered yeast. Dae-Kyun Ro et al. Apr 13, 2006, Copyright © 2006, Springer Nature (b) © SCOTT SINKLIER / AGSTOCKUSA / SCIENCE PHOTO LIBRARY (c) Courtesy of Amyris, Inc. All rights reserved

using *S. cerevisiae* engineered to produce farnesene, a hydrocarbon and excellent industrial feedstock. Despite observing high production rates in cells cultured under laboratory conditions, the company was unable to replicate this when growing strains at large scale. The company predicted production targets of 9 million litres in 2011 and 44–50 million litres of hydrocarbons in 2012, which was later reduced to 1–2 million litres in 2011. Even this lower target was not achieved. The company has scaled back plans to produce low-value, high-volume products such as farnesene and has since sold the plant, which is shown in Figure 6.9(c). The company did not put forward the exact reasons for these production difficulties. However, when organisms are engineered so that their metabolism is diverted towards high levels of production of a single metabolite and away from growth, a significant drop in fitness is expected. Therefore, any contaminant with fast growth may quickly dominant a culture, especially when a ready energy source, such as sugar, is available. It should be added that the company has received over a billion US dollars in investment and is yet to make a profit. Other biofuel companies such as Joule Unlimited, KiOR, and Cello Energy have gone bankrupt despite receiving hundreds of millions of dollars in investment. Future success in this area will be dependent on developing strains which demonstrate production of the desired metabolite over long periods, optimizing scale-up and reducing the cost of industrial plants.

Mutation of strains is not as great an issue in smaller bioreactors run for short periods. However, in order to make these processes economical, the compounds have to be of very high value. For this reason most synthetic biology companies are focusing on developing processes for production of materials, pharmaceuticals, and speciality chemicals such as fragrances and artificial sweeteners, or are engineering microbial strains that have useful properties, e.g. nitrogen fixation or bioremediation. One area which is showing great promise is development of strains to produce spider silk. Spider silk can be stronger than steel or Kevlar, but farming of spiders is not possible due to their territorial and cannibalistic nature. Therefore, many companies are trying to engineer yeast or bacteria to produce silk proteins and then spin these into commercially useful fibres. Some products have been made from spider silk, which are shown in Case study 6.2. However, no company has yet been able to mass manufacture synthetic spider silk so these products are not widespread. Despite these early issues in the field, synthetic biology companies raised over 1.4 billion USD dollars of investment in 2017, suggesting that investors still envisage a bright future for this burgeoning technology.

Case study 6.2

Synthetic spider silk—a future wonder material?

Synthetic spider silk is produced by expressing the relevant proteins in microbes. The proteins can then be spun into fibres which have been used in various products as shown in the Figure CS 6.2.

What would be some possible mechanisms for reducing the cost of mass produced spider silk? What would be some of the more useful products that could be manufactured from high-strength spider silk?

Figure CS 6.2 (a) An Adidas shoe partially constructed from spider silk; (b) the Bolt Threads Microsilk™ tie, woven from spider silk.

(a) (b)

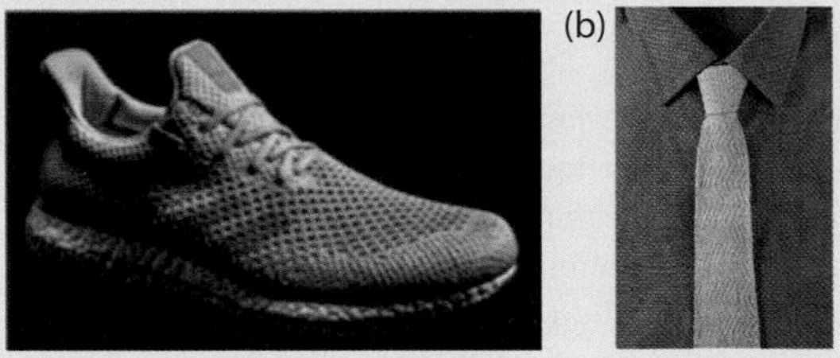

(b) Bolt Threads

Key points

- Commercializing synthetic biology is still challenging.
- Most companies are focusing on engineering strains for the production of high-value products.

Chapter Summary

- Synthetic biology is a poorly defined field which encompasses the engineering of biological organisms for useful applications.
- Synthetic biology is ideally performed in organisms that can be repeatedly genetically manipulated.
- A thorough knowledge of metabolic networks is required.
- Flux balance analysis can be used to model the effect of modifying the organism, environment or nutrient inputs.
- Gene synthesis can be used to construct novel sequences of DNA.

- Biobricks or Gibson cloning can be used to rapidly assemble DNA fragments together in a plasmid.
- Directed evolution can be used to simultaneously mutate different regions of the chromosome in selected bacterial species, thus accelerating construction of recombinant strains.
- Protein scaffolds can be utilized to alter the stoichiometry/spatial dynamics of enzymes in a metabolic pathway to optimize compound production.
- Protein microcompartments are potential mechanisms for co-localizing enzymes in a metabolic pathway to optimize compound production.
- Synthetic cells consist of a synthesized chromosome incorporating only the genes of interest and could be a chassis for specific functions, e.g. compound production.
- Economics and engineering factors are key for commercializing synthetic biology.
- Most companies are focusing on production of high-value compounds or strains with specialized applications.

Further Reading

Orth, J. D., Thiele, I., and Palsson, B. O. (2010). 'What is flux balance analysis?' Nature Biotechnology 28: 245–8.

The following is an excellent review which discusses the methods and issues of this technique.

Discussion Questions

6.1 List products that could be commercially produced using engineered microbes?

6.2 What are the ethical considerations of synthetic biology?

7 DIAGNOSTICS

Learning Objectives

- To be able to explain the basic principles of an assay;
- to be able to explain the difference between specificity and sensitivity, and why both are necessary;
- to be able to give an overview of detectable interactions used in assays;
- to be able to explain the basic parts of biosensors;
- to be able to describe cell surface display options;
- to be able to explain the nature of point of care tests and discuss their role;
- to be able to describe what an electric nose and a nano-cantilever array is.

The topic of diagnostics is exciting and complex, and we are seeing constant technological development and refinement in this field. In this chapter we will explore some of the main principles in the context of microbial biotechnology:

- using microorganisms for diagnostics;
- producing tools for diagnostics such as antibodies, enzymes, matrix components, and linkers;
- conducting diagnostics when monitoring biotechnological processes such as fermentations and bioremediation.

The focus of this chapter is the discussion of key examples of detection rather than the diagnostics of pathogens or providing an overview of all approaches to diagnostics.

There is a common framework of steps from sample (containing the **analyte** to detect) to diagnostic read-out, regardless of whether the analyte is a compound in the environment (e.g. pollutants or natural resources), in food or in forensic samples, for veterinary, agricultural, marine, and defence applications, or to detect disease markers (e.g. enzyme activity, mutations, hormones or signal transmitters). The steps and principles have changed over time with technological advancement. Diagnostics also depend on the sampling context, e.g. 'Is the sample site near a well-equipped laboratory?' or 'Is the available time limited, or the budget?'

The more we know about diagnostic principles, the better compromises we make for the benefit of clients, patients, and society.

Specificity and sensitivity are the crucial criteria of any diagnostic approach. The method of detection needs to be specific so that it can reliably and correctly identify the analyte, as well as being sensitive enough so that analytes can be detected at low concentrations.

> Frequently we have to compromise in the diagnostic approach just like when developing other biotechnological processes, such as selecting a host or expression system (remember what you learnt in Chapter 1).

7.1 Principles of diagnostics

General diagnostic approaches have frequently been driven by developments in pathogen detection, which were historically based on cultivation then moved to serological analysis, then to nucleic acid detection and then away from it again.

Antibodies, as illustrated in Figure 7.1, are the key to protein–protein-based detection, initially as the detected molecule (serology), and now as the detecting

Figure 7.1 The domains of a full-size IgG molecule and as separate functional entities. The colours allow tracing as to which part of the full antibody gives rise to the functional subfragments. The green regions facilitate contact to the antigen, whereas the blue Fc part has biological functions such as binding to phagocytes

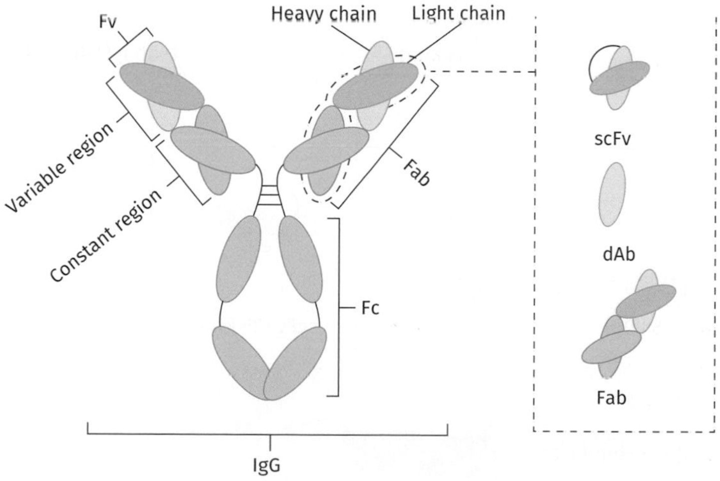

tool in the much broader context of diagnostics, where large amounts of specific antibodies are being used. Such high demand is being met by advancements in biotechnology. In 1975, hybridoma cell lines were developed to produce **monoclonal antibodies** (all these antibodies have the same specificity). Meanwhile a broad range of **recombinant antibodies** can be produced as well as their subfragments (see Figure 7.2) such as single-chain Fv antibodies (scFv is a heterodimer of the VH and VL domains connected by a linker and includes a complete antigen binding site). These can be customized for the application in question. This recombinant protein production does not only use microbial and mammalian cell cultures, but also plants (**plantibodies**). The versatility and significance of antibody-based diagnostics, as explored later in this chapter, is illustrated by the broad range of signal output possible in **immunodetection** as depicted in Figure 7.3.

Nucleic acid-based detection systems are more sensitive than antibody-based ones when the process involves amplification of nucleic acids using, e.g., polymerase chain reaction (PCR). Multiplex approaches for testing a number of

Figure 7.2 Full size IgG molecule and a range of recombinant fragments thereof. The colours allow tracing as to which part of the full antibody gives rise to the functional subfragments. The green regions facilitate contact to the antigen, whereas the blue parts have biological and stabilizing functions.

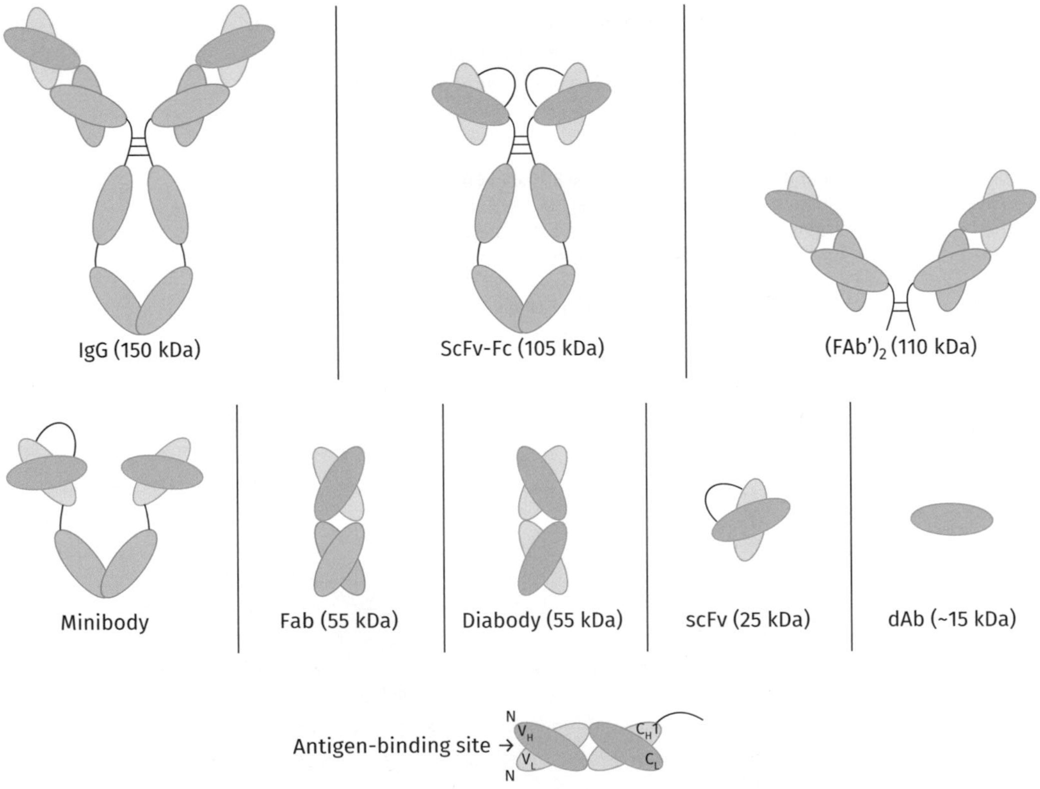

Figure 7.3 The type of measurable signal output in immunodetection can be selected and optimized. Antibodies (yellow) bind to the antigen (blue) in question and in turn are bound to by a second antibody (green) conjugated to an enzyme (orange) that catalyses a chemical reaction. The chemical reaction produces a colour change or emits light, or is labelled with a fluorophore, radioisotope (blue star) or immunogold (Au).

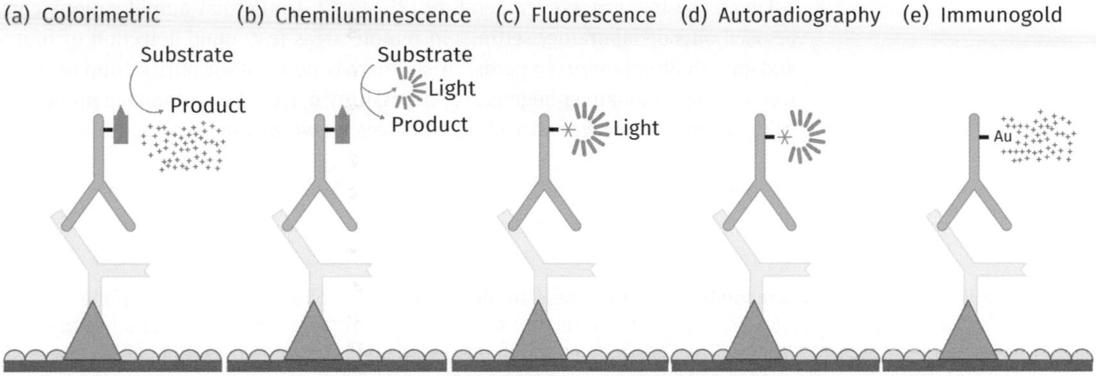

Adapted courtesy of Bio-Rad Laboratories, Inc.,© 2020

targets simultaneously are also possible. However, amplification is not always necessary, e.g. the human papilloma virus genomic DNA can be directly detected using piezoelectric biosensors.

Even if amplification is performed, it is no longer restricted to polymerase chain reactions. Isothermal alternatives such as recombinase polymerase amplification (introduced in 2006) exist. The usual heat denaturation step of the PCR is replaced by the action of *Escherichia coli* RecA recombinase and single-strand

Scientific approach panel 7.1
Pushing the (detection) limit

Commonly, biosensors are based on immobilized DNA probes which have low detection sensitivity given DNA probes of approximately 20 bases cannot usually hybridize to long target sequences. Thus, target sequences are amplified by PCR beforehand, which is time-consuming and limits broader applications. Using bis-peptide nucleic acids (a synthetic peptide backbone replaces the sugar phosphate backbone) as probes improves this. Peptide nucleic acids bind to adenine and guanine in double-stranded target DNA via strand displacement, which is more specific than that of a DNA probe. These bis-peptide nucleic acids are incorporated in piezoelectric quartz biosensors, detecting a change in mass upon binding the target. Direct detection of genomic DNA is therefore achieved without PCR amplification. Such a biosensor is applied for rapid detection of human papilloma virus 18 with a clinical specificity of 97.22 percent of that using PCR amplification, and distinguishes sequences that differ in one base only. As human papilloma virus infection can result in cervical cancer, early detection of the virus is critical for starting treatment. Using this biosensor saves time, and due to its robustness, small size, and low cost, it can be applied widely in any biomedical context.

How can this principle apply to other diseases? Is this the only way to limit potential suffering from cervical cancer?

DNA binding protein, as shown in Figure 7.4. The diagnostic tool is more affordable and easy to use. Results are available within 5–20 minutes and the method is sensitive enough to detect single nucleotide polymorphisms in human cancers or in genetically modified organisms.

Given the low process temperature (25–42°C), isothermal amplifications can be used outside laboratory settings in remote areas (e.g. rapid detection of foot-and-mouth disease virus in herds). Also, there is no need for purification or agarose gels, and human pathogens are successfully detected in automated processes using sample-to-answer microfluidic devices, as we will see later in section 7.3.

Figure 7.4 The steps of a recombinase polymerase amplification cycle. Following the binding of the recombinase and prime, a DNA strand is displaced (forming a D-loop structure) and the primer binds to the target strand. The recombinase disassembles and primer extension is performed by DNA polymerase. Subsequently, the newly generated strands are the target for the next cycle.

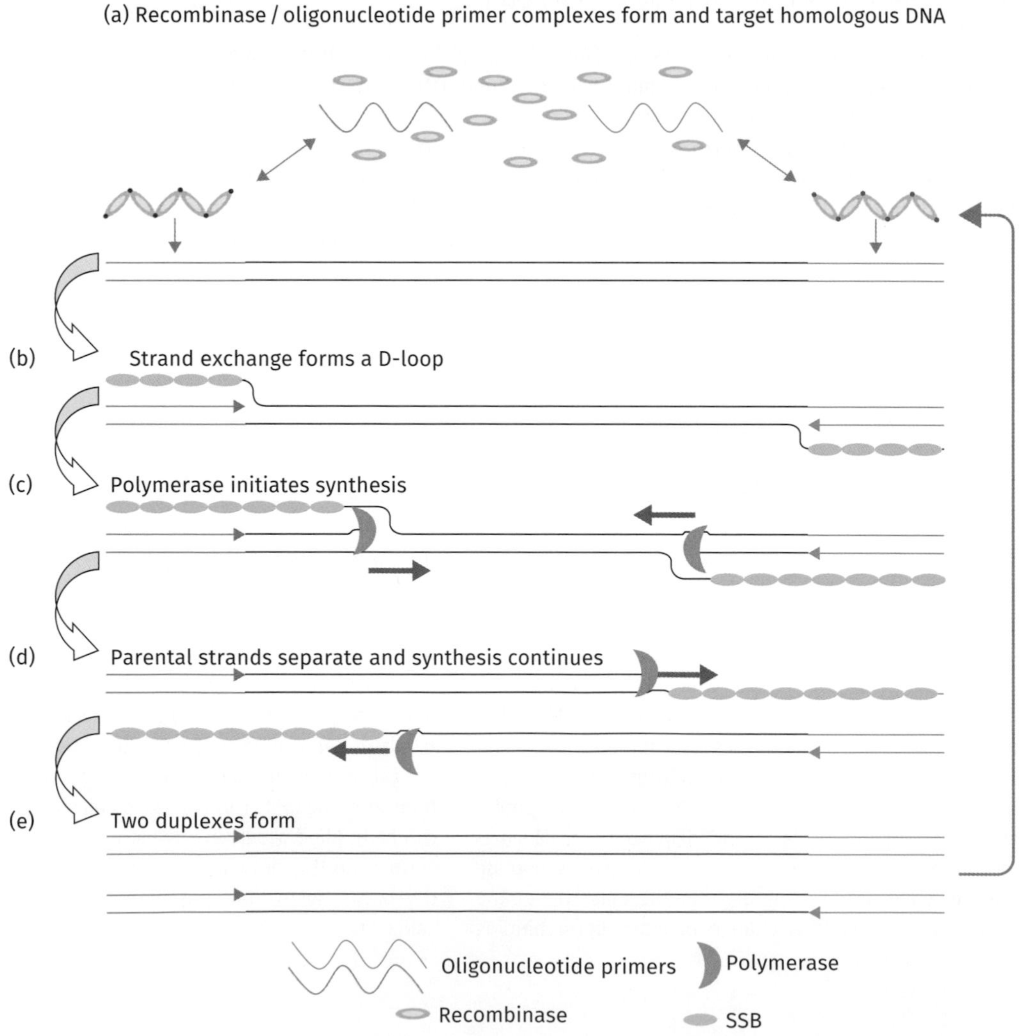
(a) Recombinase / oligonucleotide primer complexes form and target homologous DNA
(b) Strand exchange forms a D-loop
(c) Polymerase initiates synthesis
(d) Parental strands separate and synthesis continues
(e) Two duplexes form
Oligonucleotide primers
Polymerase
Recombinase
SSB

Case study 7.1
Time is of the essence

Ebola virus infections first appeared in 1976. The virus causes severe haemorrhagic fever in humans with a death rate of 50–90 percent. The largest Ebola outbreak so far took place in 2014–2016 in West Africa. There were nearly 30,000 suspected cases, half of which were laboratory-confirmed. Of these, three-quarters resulted in deaths, despite all efforts. The outbreak was eventually halted. Confirming cases rapidly with only limited resources is crucial for disease control. Reverse transcription-loop-mediated isothermal amplification allows the detection of just 20 copies of the viral RNA genome in just under half an hour. The test is even faster if the sample contains more viral particles or RNA. This is much quicker than standard methods such as real-time PCR and ELISA, and even conventional non-isothermal reverse transcription. With isothermal amplification there is no need for sophisticated equipment, highly skilled personnel or costly consumables, which makes it very well suited for, e.g., field diagnostics during Ebola outbreaks in Central Africa.

Would you recommend the use of this technology also when results are not required swiftly?

Could this technology be used for other disease outbreaks?

Key points

- Detection methods need to be specific (so that the analyte can be reliably and correctly identified) and sensitive (so that the analyte can be detected at low concentrations).
- Nucleic acid-based detection systems are more sensitive than antibody-based ones when the process involves amplification of nucleic acids.
- Recombinant antibodies and their subfragments are instrumental in antibody-based diagnostics.

7.2 Mechanisms of detection

The basis of any diagnostic development is a specific interaction. This interaction or binding event can either be detected directly or the binding triggers a subsequent output such as the expression of a reporter gene. Detectable interactions take place between;

- proteins;
- nucleic acids;
- between protein and nucleic acid.

In a biosensor these interactions are performed by the **bioreceptor**. Typically, a biosensor has three parts:

1. The bioreceptor, such as an antibody or DNA, which interacts with the analyte;
2. The **transducer element**, which converts the biological response into an electrical, optical, or thermal signal;
3. The **reader device**, which detects the signal output.

Figure 7.5 Phage displaying an scFv antibody. The phage particle with the displayed scFv is depicted.

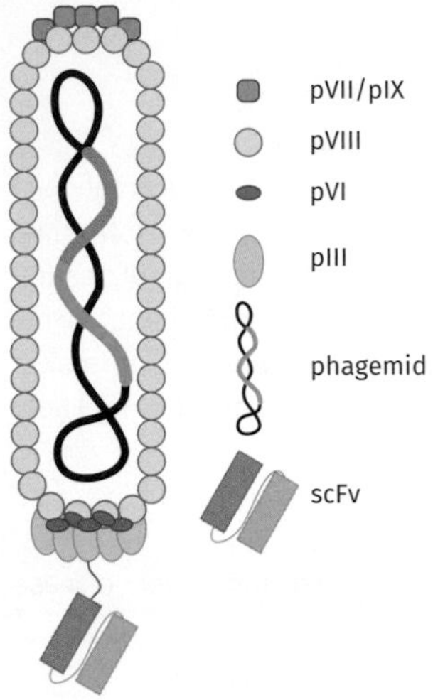

Adapted with permission from Michael Hust, Sebastian Miethe, André Frenzel, et al. Recombinant antibodies for diagnostics and therapy against pathogens and toxins generated by phage display. PROTEOMICS - CLINICAL APPLICATIONS, 21 June, 2016. John Wiley and Sons Copyright © 2016 WILEY-VCH Verlag GmbH & Co. KGaA, Weinheim.

Interactions on the cell surface and inside the cell

An elegant means of protein-based detection technology is **phage display** (see Figure 7.5), where the genetic information for an antibody fragment such as scFv is introduced into the **phage** genome in combination with a viral coat protein gene. During expression in *E. coli* the antibody fragment fused to the viral protein is synthesized, and the antibody fragment is displayed when the phage assembles. If the genetic information of the antibody fragment is fused to that of marker proteins such as fluorescent proteins or alkaline phosphatase, the labelling necessary for immunodetection takes place directly *in vivo* during recombinant production, which saves cost and time. The phages displaying such labelled antibody fragments can be tailored and optimized for the desired diagnostic assay e.g. for botulinum neurotoxin, staphylococcal enterotoxin B or biological threat agents such as *Bacillus anthracis*.

If bacterial cells are used for display instead of phages a plethora of cell surface display options becomes available as illustrated in Figure 7.6.

Interactions inside cells (see Figure 7.7) can be used instead of interactions on the cell surface. Diagnostics using this approach involve entire cells as biosensors to provide the readout. Whole-cell biosensors can be used in environmental analyses (e.g. water or soil pollution) or medical diagnostics (e.g. gastrointestinal diseases).

Key points

- The basis of any diagnostic assay is a specific interaction, which can either be detected directly or triggers a subsequent measurable output.
- Detectable interactions take place between proteins, nucleic acids, protein, and nucleic acid.
- A biosensor has three parts: the bioreceptor, the transducer element, the reader device.
- Cell surface display options for protein-based detection technology use bacterial cells or phage display.
- If intracellular interactions are used, bacterial cells are turned into whole-cell biosensors.

Interactions at the level of transcription factors and riboswitches

A key aspect when developing a whole-cell-based biosensor is choosing an appropriate **reporter gene**. The reporter gene will allow amplification of the molecular recognition when a regulator protein binds (i.e. detects) its particular target analyte. Thus the bacterial cell functions as a highly sensitive and selective recognition element for **high-throughput** *in situ* detection, as we will discuss in section 7.3. Selecting a reporter gene is a compromise depending on the diagnostic context. Table 7.1 provides an overview of several biosensor sensitivities. Bacterial luciferase gene (*lux*), firefly luciferase gene (*luc*), green fluorescent protein gene (*gfp*), and β-galactosidase gene (*lacZ*) are commonly used as reporters. While GFP does not require a substrate or ATP, it is less sensitive than luciferases and unsuitable for rapid detection, while luciferases are

Figure 7.6 Overview of applications of cell surface display technologies. The displayed components are anchored in the cell membrane and face outwards.

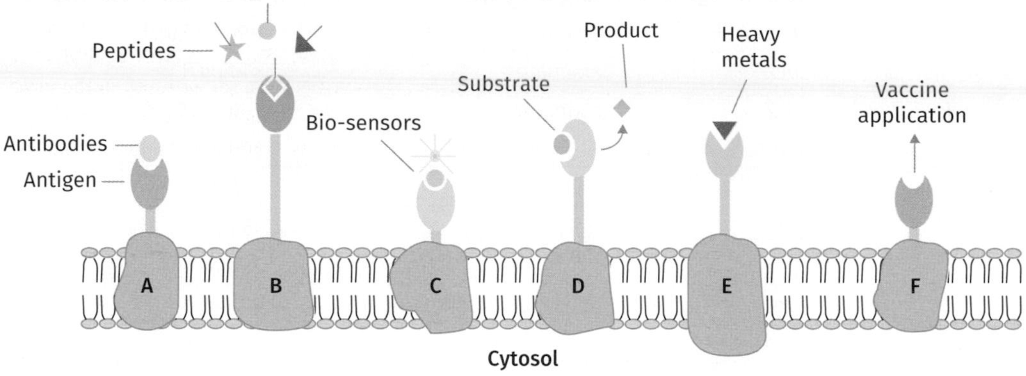

Adapted from Ravikumar, S., Baylon, M.G., Park, S.J. et al. Engineered microbial biosensors based on bacterial two-component systems as synthetic biotechnology platforms in bioremediation and biorefinery. Microb Cell Fact 16, 62 (2017) doi:10.1186/s12934-017-0675-z. Distributed under the terms of the Creative Commons Attribution 4.0 International (CC BY 4.0). https://creativecommons.org/licenses/by/4.0/.

Figure 7.7 Diagram of a whole-cell-based biosensor. The flow of information from analyte to readout is shown.

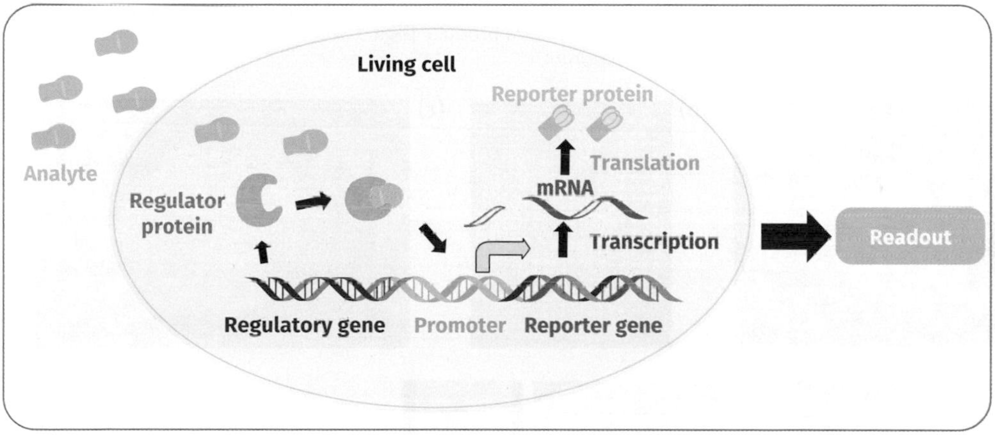

Gui, Q.; Lawson, T.; Shan, S.; Yan, L.; Liu, Y. The Application of Whole Cell-Based Biosensors for Use in Environmental Analysis and in Medical Diagnostics. Sensors 2017, 17, 1623. Distributed under the terms of the Creative Commons Attribution 4.0 International (CC BY 4.0). https://creativecommons.org/licenses/by/4.0/.

not thermostable. A sensitive and rapid solution for both colourimetric or fluorescent read-outs is *lacZ*. The carotenogenic gene *crtA* does not even require the addition of a substrate but rapidly changes the cultivation medium colour from yellow to red. Signal output can be seen in Figure 7.8.

The molecular interaction/detection can be established using transcription factors—proteins that sense changes and in response regulate gene expression. An analyte-specific promoter can be inserted in the host transcription system to trigger reporter gene expression.

Table 7.1 Comparison of whole-cell-based biosensors in their sensitivities (reproduced from Gui et al. 2017)

Host cell	Reporter gene	Target analyte	Detection sensitivity
Escherichia coli	*lux*CDABE	arsenic	0.74–69 μg/l
Escherichia coli	*lacZ*	arsenate	<10 μg/l
Deinococcus radiodurans	*lacZ*	cadmium	1–10 mM
	crtI		50 nM–1 mM
Escherichia coli	*gap*	chromate	100 nM
Escherichia coli	*gfp*	zinc, copper	16 μM
			26 μM
Escherichia coli	*luc*	benzene, toluene, xylene	40 μM
Escherichia coli	*lux*AB		0.24 μM
Pseudomonas putida	*lux*AB	phenol	3 μM
Burkholderia sartisoli	*lux*AB	naphthalene, phenanthrene	0.17 μM
Escherichia coli	*lux*AB	C_6–C_{10} alkanes	10 nM
Escherichia coli	*lux*CDABE	tetracyclines	45 nM

Figure 7.8 Signal output for a range of reporter genes. (a) Recombinant *E. coli* containing the *lux* gene emit light; (b) recombinant *E. coli* containing the bioluminescent protein firefly luciferase; (c) recombinant *E. coli* containing the green fluorescent protein; (d) blue-coloured recombinant bacteria produce β-galactosidase transforming the substrate X-gal; (e) carotenoid production in *Fusarium oxysporum*.

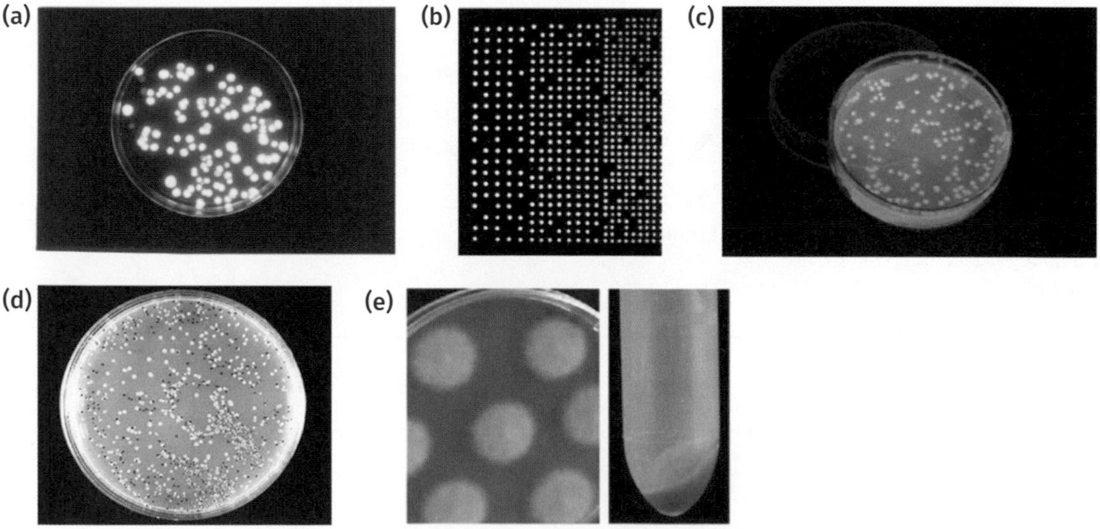

Alternatively, two-component systems (see Figure 7.9) can be exploited for detection purposes. A histidine kinase acts as sensor and, following phosphorylation of the response regulator by the sensor, the response regulator drives effector gene expression. These systems can be engineered for different diagnostic applications even as chimeric two-component systems to detect novel compounds.

Engineered microbial biosensors can be used as synthetic biotechnology platforms for bioremediation of contaminated soil and groundwater, biorefinery, and bio-adsorption of heavy metals. If the microorganisms are designed to produce extracellular enzymes or display enzymes on their surface, they can act as whole-cell catalysts. Applications for the production of, e.g., biofuels, fine chemicals, and polymers from renewable resources require combining biosensor and transforming/synthetic abilities in one cell. In either scenario, the bacterial system behaves normally until the two-component regulatory system detects the target in the environment.

Riboswitches are a further molecular interaction mechanism that can be utilized for detection purposes. They are regulatory elements commonly situated in the 5′-untranslated region of mRNAs, as illustrated in Figure 7.10. The downstream gene expression is controlled by direct binding of the aptamer domain of the riboswitch to a small molecule ligand, which is the effector molecule that is being sensed/detected. Due to this binding, the riboswitch changes its structure and this is how transcription is regulated.

Whole-cell biosensors based on a riboswitch respond fast, and can be engineered for various applications, e.g. environmental monitoring.

Depending on whether a transcriptional repressor or activator is used when engineering the switch, the readout can be tailored. There are even inverter switches that produce a reciprocal response. Biphasic switches can use negative and positive regulation to sensitively respond to low analyte concentrations. Toggle switches have two repressors that cross-regulate each other's downstream elements, and similar to electronic circuits (see Figure 7.11), a range of logic gate types such as 'NOR', 'NAND', and 'NOT' have been developed for biological circuits in biosensors.

Figure 7.9 A general diagram of a two-component regulatory system. The flow of information from signal to effector gene transcription is shown.

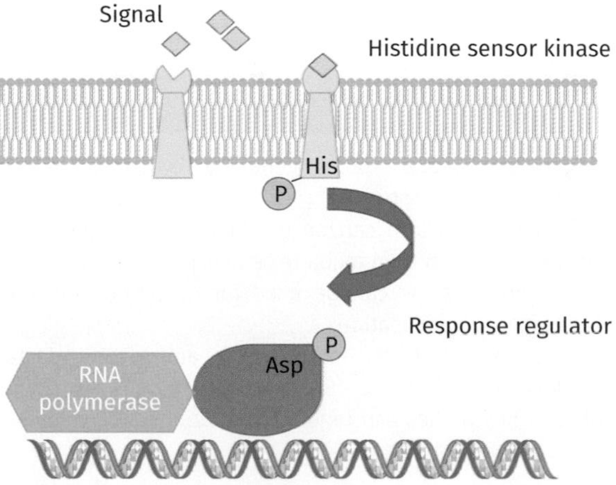

Figure 7.10 Diagram of a riboswitch. The elements from 5′ to 3′ are indicated.

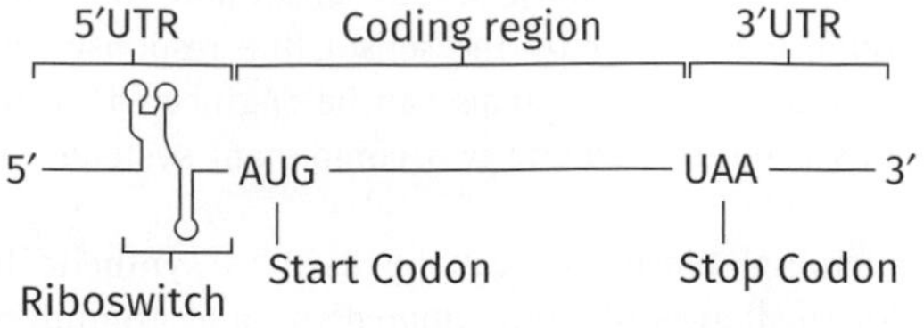

Edwards, A. L. and Batey, R. T. (2010) Riboswitches: A Common RNA Regulatory Element. Nature Education 3(9):9. © 2010 Nature Education.

Figure 7.11 Logic gates as used for electronic circuits. Based on the structure of the gate, input A and B are transformed into output X.

Name	NOT	AND	NAND	OR	NOR
Symbol	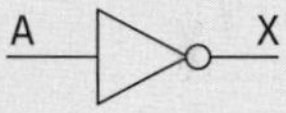	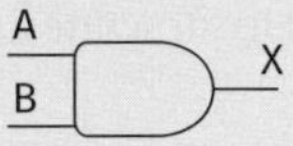	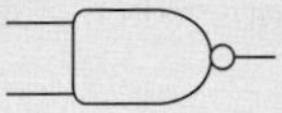	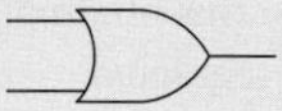	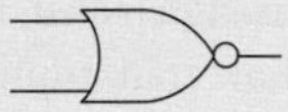
Truth Table	A X 0 **1** 1 **0**	B A X 0 0 **0** 0 1 **0** 1 0 **0** 1 1 **1**	B A X 0 0 **1** 0 1 **1** 1 0 **1** 1 1 **0**	B A X 0 0 **0** 0 1 **1** 1 0 **1** 1 1 **1**	B A X 0 0 **1** 0 1 **0** 1 0 **0** 1 1 **0**

Key points

- Choosing an appropriate reporter gene is a key aspect when developing a whole cell-based biosensor.
- The molecular interaction can be established using transcription factors, sensing changes which then regulate gene expression.
- Two-component systems can be used for detection, and engineered for different diagnostic applications.
- Riboswitches can be utilized for detection, and respond faster than transcription factors.
- Engineered riboswitches can regulate outputs like logic gates in electronic circuits.

7.3 Biosensors and their applications

Biosensors (see Figure 7.12) are categorized depending on the bioreceptor type detecting the analyte, or their transduction mode (optical, piezo-electric, or electrochemical) converting the biological response into a signal. Due to technological advances in biotechnology and microelectronics, biosensors have become increasingly sophisticated. The fastest growing technology in pathogen detection is biosensors. Biosensors are highly specific, fast, only require limited reagents, and even detect and measure non-polar compounds. Combining biosensors and nanotechnology (1–100 nanometres), allows further progress in detection and quantification, even remotely off site.

Figure 7.13 shows an electrochemical biosensor. Its cost and specificity depends on the type of bioreceptor such as aptamers, enzymes, antibodies, or bacteriophages. The measured output is a change in electronic current, or a conductance change carried by bioelectrodes.

Figure 7.12 Diagram of a typical biosensor. The flow of information from analyte to readout is shown.

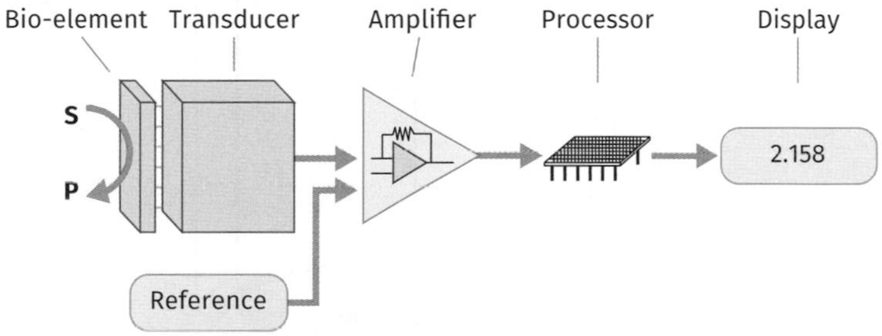

RF Wireless World

Figure 7.13 Diagram of an electrochemical biosensor. The flow of information from analyte to readout is shown.

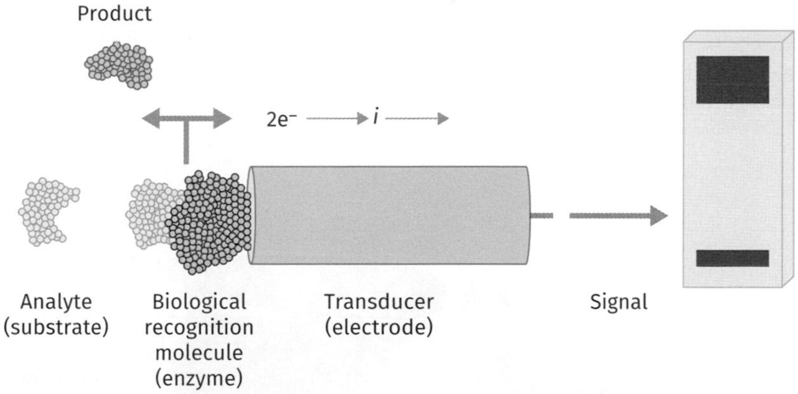

Adapted from Putzbach, W. et al., Immobilization Techniques in the Fabrication of Nanomaterial-Based Electrochemical Biosensors: A Review, Sensors 2013, 13, 4811–4840. This article is an open access article distributed under the terms and conditions of the Creative Commons Attribution license (http://creativecommons.org/licenses/by/3.0/).

Amperometric biosensors are based on the movement of electrons (i.e. a current) produced by an enzymatically catalysed redox reaction of the analyte, as shown in Figure 7.14. The electronic current is measured, and analyte concentration is determined due to the directly proportional relationship to the current.

A good example of an amperometric biosensor is the glucose biosensor used by diabetes patients to monitor blood glucose levels (see Figure 7.15) or in the fermentation industry to measure glucose concentrations in bioprocesses (remember Chapter 1). Glucose oxidase catalyses the redox reaction of glucose and oxygen to gluconic acid and water. The oxidation releases electrons, which are accepted

Figure 7.14 Diagram of an amperometric biosensor for the detection of glucose. The translation of glucose presence into a current is shown.

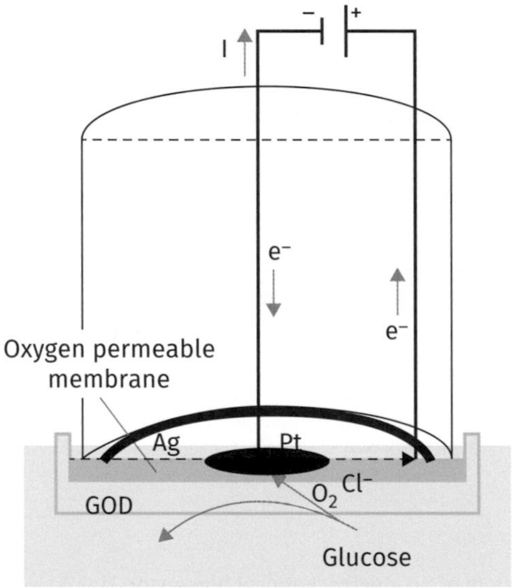

Emeritus Professor Martin Chaplin London South Bank University

Figure 7.15 Example of a blood glucose biosensor. The glucose presence in the blood is translated into a readout.

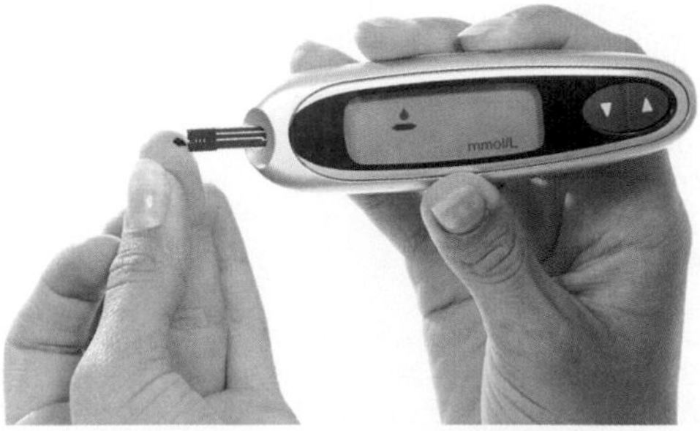

Dmitry Lobanov/Shutterstock.com

during reduction as shown in Figure 7.14. The sensor has a central platinum cathode and a surrounding silver/silver chloride anode situated in a potassium chloride electrolyte solution. *Penicillium* spp. or *Aspergillus* spp. recombinantly produce glucose oxidase.

Piezoelectric (or acoustic) biosensors as depicted in Figure 7.16 measure signal output based on frequencies. The piezoelectric crystals (e.g. quartz) in the sensor vibrate at a specific resonant frequency when in an electric field. The change in resonant frequency if analytes bind to the bioreceptor (the antibody), is measured.

Figure 7.16 Diagram of a piezoelectric biosensor. The elements of the sensor are shown. The antigens are being detected.

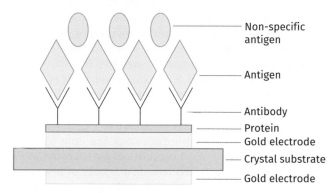

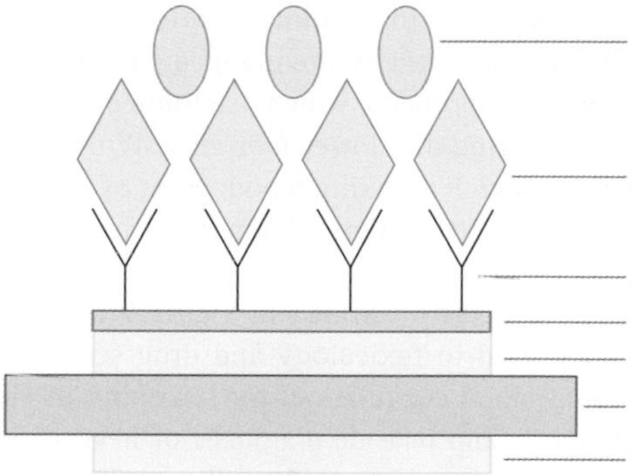

Key points

- Biosensors are categorized based on the type of bioreceptor, or their transduction mode (optical, piezo-electric, or electrochemical).
- Combining biosensors and nanotechnology allows further progress in detection and quantification, even remotely, off site.
- The output of electrochemical biosensors is a change in electronic current or conductance.
- Amperometric biosensors are based on a current produced by an enzymatically catalysed redox reaction of the analyte.
- Piezoelectric (acoustic) biosensors measure signal output based on frequencies.

Point of care tests

The technological progress in diagnostic methods and equipment, and achieving read-outs in real or near real time, has allowed for the development of point of care tests. These tests move molecular diagnostics from central laboratories, with expensive and complex high-precision instruments, to wherever the patient is located. Nanomaterials enable previously unachievable detection of extremely small changes in analyte concentration. The use of microfluidics (liquid flowing in channels where at least one dimension is less than 1 millimetre), isothermal amplification reactions, miniaturized detectors, cheap imaging technologies, and wireless signal transmission made it possible to build 'sample in–results out' diagnostic devices. To aid in the development of such test devices the World Health Organization has defined criteria that all point of care diagnostic tests should meet. These criteria are referred to by the acronym ASSURED, which stands for: affordable, sensitive, specific, user-friendly, rapid and robust, equipment-free, and deliverable to those who need it. Such point of care tests (e.g. for diagnosing malaria, tuberculosis, HIV infections) are invaluable in support of disease monitoring and control measures in the developing world, and disadvantaged populations in the developed world.

The self-contained and portable point of care test devices can be used, without training, to detect pathogens. Routine patient screening and identifying therapeutic concepts can be easily facilitated, because patient response to treatment can be monitored. Gaining results quickly in non-laboratory settings allows a more patient-centred health care. The read-outs can be kept as an electronic medical record and immediately shared to coordinate patient care. The role of point of care tests cannot be overestimated in moving medicine from a curative to a personalized and pre-emptive approach, which use limited resources more efficiently and result in less suffering and lower mortality. Without having to use invasive sampling procedures such as taking blood, point of care diagnostic tests can also detect patient-generated airborne biomarkers found in aerosols and volatile compounds from breath and odour.

Aside from detecting pathogens (as illustrated in Figure 7.17) or biothreat agents, point of care tests are used in toxicology and drug screens, for haematological analysis including blood coagulation, for screening of circulating tumour cells, in stroke diagnosis, and bedside diagnosis of heart disease. It is expected that in 2020 globally, approximately 20 billion USD will be spent annually on point of care tests. Robust standards for point of care test data still need to be set for quality assurance and control requirements.

The bigger picture panel 7.1
Where it counts

Infectious diseases have an undeniable socio-economic impact. They have frequently changed history in its course—the destruction of the feudal system in Europe in the wake of the fourteenth-century plague epidemics when two-thirds of the European population died, is one such example. In countries with low gross domestic product, the budget for health care, disease control and prevention, including educational initiatives is limited. This is a vicious cycle given infectious disease and related loss of labour lowers income further. As an example, average life expectancy is shortened by about an eighth for an individual with an HIV infection without access to treatment. The growth rate of the gross domestic product is reduced per capita per year by about 1.5 percent in regions with an HIV prevalence of about 30 percent if there is little or no treatment available. In this context HIV/AIDS weakens agriculture, industry, governance, civil service, and armed forces. In a household with one AIDS death, the cotton and vegetable production is halved. If women are ill, girls take over family duties and miss school, which means childcare, nutrition, and education suffer. Early treatment initiation based on rapid diagnosis in low-cost settings can be achieved with point of care tests for diagnosing HIV infections.

What are the issues in this complex situation, and are there any that have not yet been alluded to? What are the challenges in addressing any of the issues?

Figure 7.17 **Overview of target products, users, and settings of point of care testing.** HBV—Hepatitis B Virus; HCV—Hepatitis C Virus; UTI—urinary tract infection; MRSA—methicillin-resistant *Staphylococcus aureus*; C. diff—*Clostridium difficile*; RDT—rapid diagnostic test; Strep A—group A *Streptococcus*.

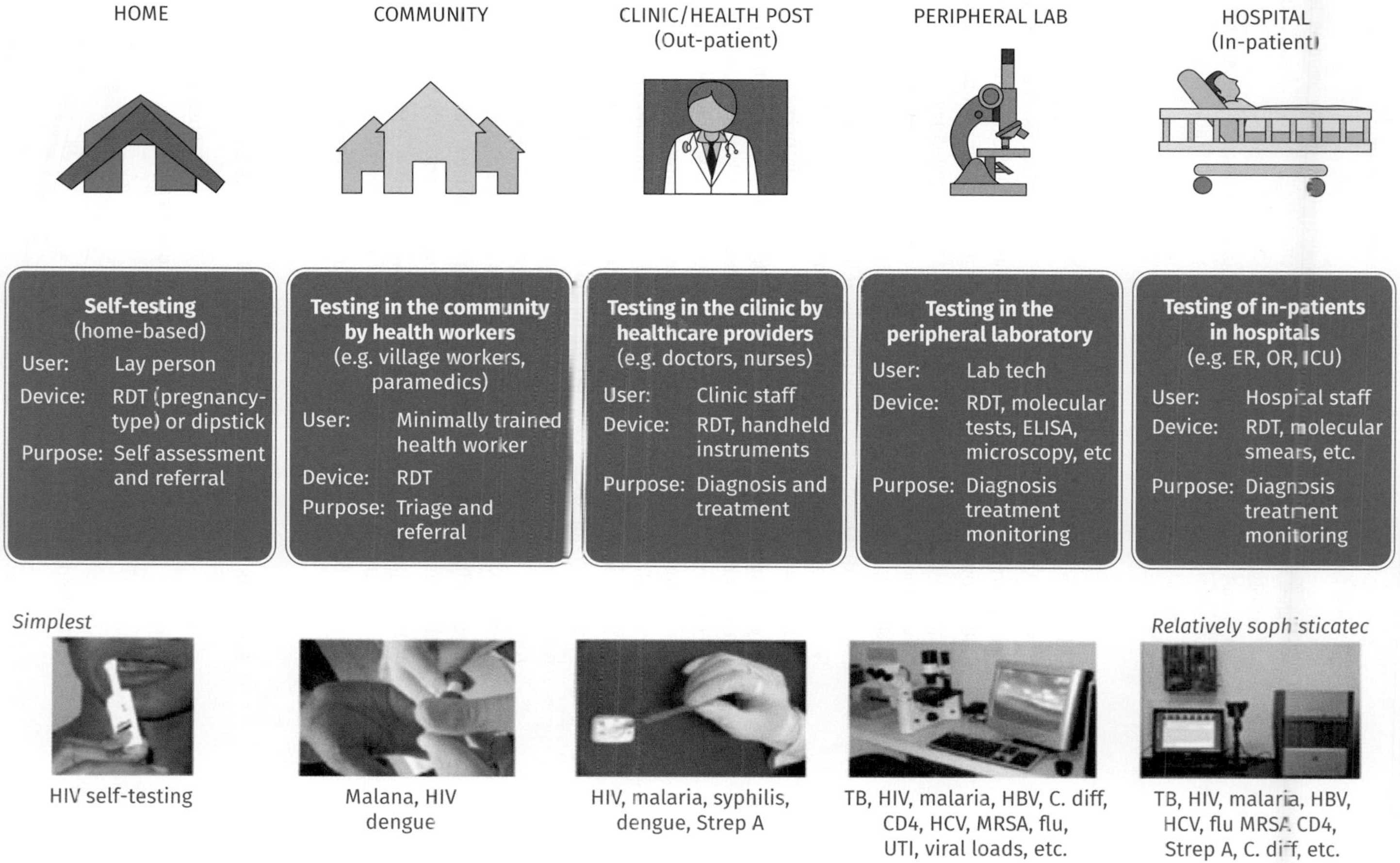

Adapted from Pai NP, Vadnais C, Denkinger C, Engel N, Pai M (2012) Point-of-Care Testing for Infectious Diseases: Diversity, Complexity, and Barriers in Low- And Middle-Income Countries. PLoS Med 9(9): e1001306. https://doi.org/10.1371/journal.pmed.1001306.

Key points

- Point of care tests move molecular diagnostics from central laboratories to the patient.
- The use of microfluidics, miniaturized detectors, cheap imaging technologies, and wireless signal transmission allow for 'sample in–results out' diagnostic devices.
- Point of care tests are invaluable in support of disease monitoring and control measures in the developing world and disadvantaged populations in the developed world.

Lateral flow tests

A successful example of point of care diagnostics is the lateral flow test (see Figure 7.18). Its development dates back to 1956 and is based on latex agglutination tests. Nowadays it also detects proteomic or genomic biomarkers. The output is qualitative (yes/no) as in the case of pregnancy test kits, or quantitative, and can be multiplexed to detect several analytes in one approach. No washing steps are required but with an analytical sensitivity in the micromolar range the diagnostic use for pathogens at low concentrations is limited. The analyte separation and detection takes place on a matrix enabling the flow of liquid such as urine or blood driven by **capillary action**. Matrices can be nitrocellulose, polymer or paper, as used when measuring glucose. As shown in Figure 7.19, a labelled antibody is pre-loaded on the matrix and binds the analyte contained in the sample. The bound complex migrates on the strip until it binds a further target-specific antibody. The capture antibody is immobilized on the matrix, and the complex does not travel any further. Upon accumulation of captured labelled antibody-analyte complexes it all becomes visible as a line. The line is either detected by the user's eye or can be measured by a detector. As a control of test function, a non-specific antibody is used, resulting in a separate visible line. Aptamers and molecularly imprinted polymers are capture probes alternative to antibodies. Latex beads, colloidal gold or carbon, enzymes, fluorescent tags, or streptavidin are used for labelling. Gold nanoparticles have a higher extinction coefficient than organic dyes, and therefore are frequently used for labelling in commercial lateral flow tests. Such rapid lateral flow tests in a dipstick format are available in developing countries for pathogens including HIV-1 and HIV-2, Hepatitis B virus, and *Mycobacterium tuberculosis*. The biotech sector is crucial

Figure 7.18 Diagram of a lateral flow test. All elements are labelled.

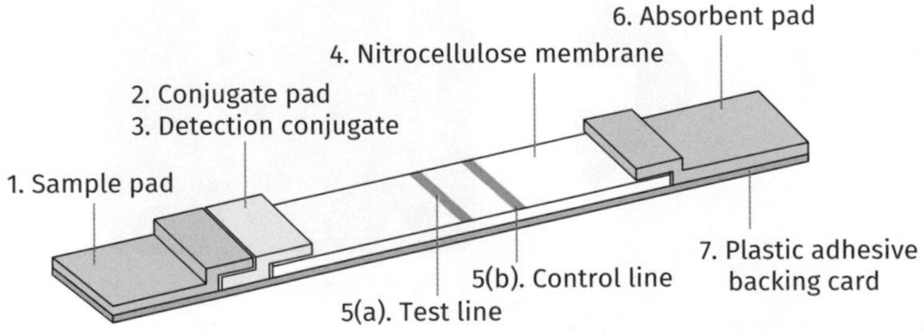

Figure 7.19 Principle of the immunoassay in context of a lateral flow test. All elements are labelled. The conjugated antibodies bind to the analyte and migrate to the test line where they bind and produce a readout. Unbound conjugated antibodies travel to the control line where they bind and produce a readout.

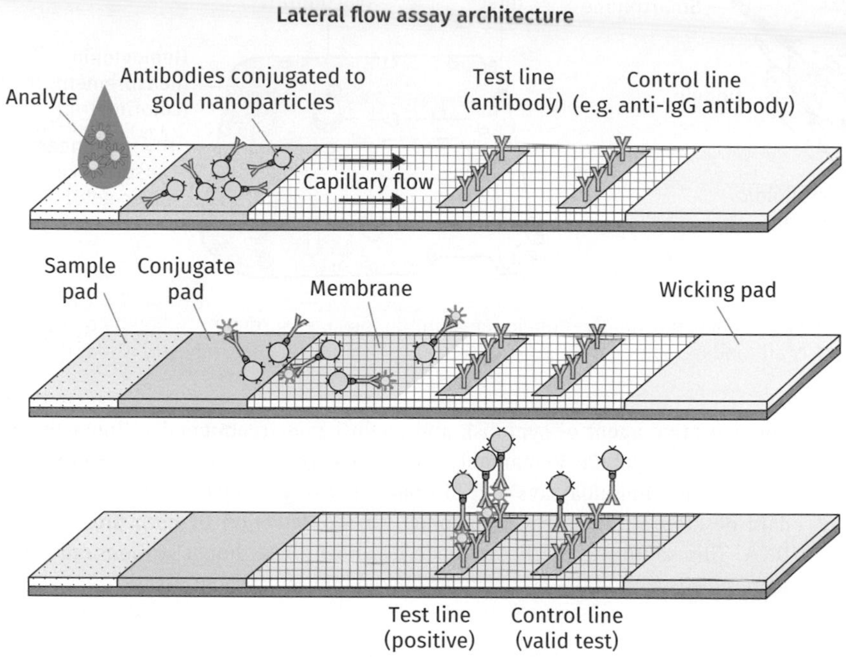

for not only developing the assaying technology, but in producing flow matrices, labelling components, and recombinant antibodies.

Even integrating two types of diagnostic assays is possible. A cheap and rapid microfluidic-based smartphone dongle (see Figure 7.20) for antenatal care combines a colourimetric quantitative assay of the haemoglobin concentration and an HIV antibody immunoassay using a spectrophotometric readout. This dual panel is used in clinical settings for diagnosis of HIV and anaemia for timely and effective treatment. The diagnostic data of many patients can be collected via the smartphone interface and fed into patient databases.

Test processes have to be compatible to run different types of diagnostic tests on one platform, yet the different concepts of detecting the signal need to be integrated in the same hardware. This nanodiagnostics platform streamlines an entire laboratory-quality immunoassay into a smartphone accessory, and can be adapted to other diagnostic tests with a colourimetric readout. The dongle is basically used in place of an enzyme-linked immunosorbent microplate assay, and the disposable plastic cassettes are preloaded with the necessary assay components. Where enzyme-linked immunosorbent assays use substrates and enzymes, gold nanoparticles, and silver ions are instrumental here in the amplification steps. The smartphone powers the electronic, mechanical, and optical processes. Whole-blood samples obtained can be analysed for blood-borne pathogens. Even triplexed diagnostic approaches based on immunoassays are possible in a single test using antibodies against HIV, *Treponema pallidum*

Figure 7.20 Microfluidic-based smartphone dongle. (a) Dongle with smartphone; (b) microfluidic cassette (test zone 1: haemoglobin measurement, test zones 2–4: HIV immunoassay with controls).

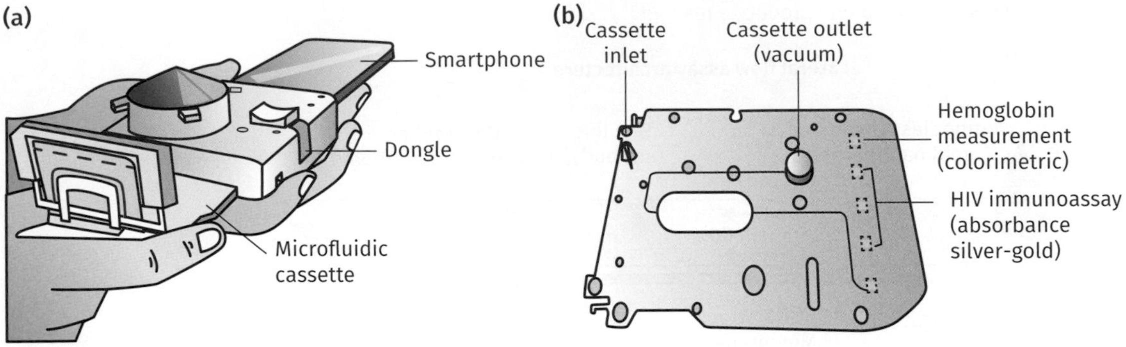

Reproduced with permission from Guo et al., Smartphone dongle for simultaneous measurement of hemoglobin concentration and detection of HIV antibodies, Lab Chip 15, 3514. 2015-07-09. Royal Society of Chemistry (Great Britain)

(the causative agent of syphilis), and against non-treponemal epitopes reflective of active syphilis. Remarkably, the results are available within 15 minutes.

Magnetic immunoassays and bio-barcode assays that can be read by standard desktop scanners can be applied for the detection of both proteins and DNA. This sounds like a home-made diagnostic test, but is a thousand times more sensitive than the best enzyme-linked immunosorbent assay.

Key points

- The output of lateral flow tests can be qualitative or quantitative and can be multiplexed to detect several analytes in one approach.
- The separation and detection of the analyte or several analytes is driven by capillary action.
- The integration of two types of diagnostic assays is possible.

Whole-cell assays

Whole-cell assays are more tolerant than isolated bioreceptors to, and therefore functional in, a wide range of environmental conditions such as pH, temperatures, and ionic strengths. This is enhanced if extremophile microorganisms (e.g. halophiles, alkaliphiles, thermophiles) are used as host cells for the biosensors. An example is the protozoan *Tetrahymena thermophila* for the detection of heavy metal ions with a luciferase output. In this way detection and measurements can be conducted remotely in extreme environments in real time.

› You will find out more about the uses of extremophiles in Chapter 9.

Detection is highly specific, sensitive, rapid, and low cost. There is little sample preparation required, making whole-cell assays ideal platforms for high-throughput screening. These systems can also be automated and miniaturized.

The variety of suitable and available matrices (e.g. silica, gold, paper, microfluidic platforms, microarrays, optical fibres, integrated circuit chips) for portable whole-cell biosensors allows flexible integration in existing diagnostic systems with optimized readouts. The whole cells can be immobilized and protected on the matrices using agarose, alginates, or agar medium.

Strip whole-cell biosensors are used, e.g. to detect bacterial quorum sensing signalling molecules (N-acyl homoserine lactones for Gram-negative bacteria, oligopeptides for Gram-positive bacteria). (For a deeper understanding of this inter-cell communication read the Microbial Physiology primer in this series.)

Escherichia coli cells have for example been engineered as biosensors by combining the *Pseudomonas aeruginosa* quorum-sensing regulatory system as a recognition element with β-galactosidase as the reporter protein. If N-acyl homoserine lactones at nanomolar concentrations are present in patient samples due to an infection, this can be monitored. Moreover, it is possible to screen for molecules that interfere with quorum sensing such as **agonists** and **antagonists**, which can then be developed for use in therapy or industry.

Using the positive feedback loops from quorum-sensing systems is also an economic and environmentally friendly approach to detect specific metals such as mercury, e.g. with whole *E. coli* cell biosensors that had a mercury-specific operon inserted. Traditional chemical and physical detection methods require expensive equipment, and the materials used for the complicated procedures frequently become toxic waste. Wastewater quality can be monitored reliably this way.

Whole-cell immunosensors with electrochemical readout have been developed by engineering yeast cells to surface display both gold-binding peptides and single chain variable fragment antibodies for the detection of specific antigens in samples.

Using naturally bioluminescent (e.g. *Vibrio fischeri*) or genetically engineered bioluminescent bacteria creates portable optical biosensors that may be used for real-time online detection of pollutants, for example.

Electronic noses

Electronic nose techniques are used for non-invasive monitoring by detecting volatile organic analytes directly in samples. The detection is based on whole cells or proteins, and is inspired by the principles of the mammalian olfactory system. Such noses (and similar tongues) are applied in the food sector to assess quality and safety, nutritional value and freshness, smell, flavour, and texture. Microbial fermentations are monitored using electronic noses to rapidly and reliably detect any irregularities in the bioprocess to assure quality. Electronic noses also detect volatile disease markers in clinical material. Various environmental (e.g. water, wastewater) and industrial (e.g. beverages) analyses are conducted using polypyrrole-based electronic noses for detecting toxic and non-toxic substances (e.g. oxygen, hydrogen, nitrogen oxides, ammonia, hydrogen sulphide, sulphur dioxide, carbon monoxide, alcohols, methane, phenol, and benzene).

Key points

- Whole-cell assays (particularly extremophiles) have a better tolerance than isolated biosensors to a wide range of environmental conditions such as pH, temperatures, and ionic strengths.
- Biosensor detection can be based on the bacterial quorum-sensing system.

- Naturally bioluminescent microorganisms or genetically engineered bioluminescent bacteria can be used in portable optical biosensors.
- Electronic noses are used for non-invasive monitoring by detecting volatile organic analytes directly in samples.

7.4 Detection of microbial cells

Rapid and accurate detection of pathogens is possible directly without any labelling when using **bacteriophages** as specific receptors. One example is the T2 bacteriophage biosensor that detects the *Escherichia coli* B **strain** electrochemically. The phages, acting as bioreceptors, are immobilized, e.g. on a glassy carbon electrode. The charge difference between the negative phage capsid and positive tail fibres can be used to immobilize phages on the surface of an electrode via electrostatic interactions.

If an *E. coli* B cell binds to a T2 phage on the electrode, a change in current can be detected using electrochemical impedance spectroscopy to the limit of about 100 bacterial cells per millilitre. Phages can withstand harsh environmental conditions much better than isolated bioreceptors (e.g. antibodies, aptamers), and can distinguish viable from dead cells unlike, e.g., DNA-based detection methods. Bacteriophages make for cheap, abundant, and highly specific bioreceptors for their host bacteria. It is important not to confuse this diagnostic approach with the above discussed phage display.

Detecting and identifying bacteria accurately at low concentrations in complex samples is also possible using nanocantilever arrays. Cantilevers are beams that are fixed at one end and free at the other, acting as mass sensors. Bioreceptors are immobilized on the beam, and if analytes bind to the receptors, the mass on the cantilever increases, leading to a read-out. Some nanocantilevers are sensitive enough to detect a single *E. coli* cell.

Key points

- Detection of bacteria (including distinguishing viable from dead cells) is possible directly without any labelling when using bacteriophages as receptors.
- Nanocantilever arrays detect and identify bacteria accurately at low concentrations in complex samples.

Chapter Summary

- Detection methods need to be specific (so that the analyte can be reliably and correctly identified) and sensitive (so that the analyte can be detected at low concentrations).
- The basis of any diagnostic assay is a specific interaction, which can either be detected directly or triggers a subsequent measurable output.

- Detectable interactions take place between proteins, nucleic acids, protein, and nucleic acid.
- A biosensor has three parts: the bioreceptor, the transducer element, the reader device.
- Biosensors are categorized based on the type of bioreceptor, or their transduction mode (optical, piezo-electric, or electrochemical).
- Combining biosensors and nanotechnology allows the detection and quantification of biomolecules at very low concentrations, as well as remotely, off site.
- Cell-surface display options for protein-based detection technology use bacterial cells or phage display.
- If intracellular interactions are used, bacterial cells are turned into whole-cell biosensors.
- Using naturally bioluminescent microorganisms or genetically engineered bioluminescent bacteria results in portable optical biosensors.
- Point of care tests move molecular diagnostics from central laboratories to the patient.
- The use of microfluidics, miniaturized detectors, cheap imaging technologies, and wireless signal transmission allow for 'sample in–results out' diagnostic devices.
- Point of care tests are invaluable in support of disease monitoring and control measures both in the developing world and disadvantaged populations in the developed world.
- Electronic noses are used for non-invasive monitoring by detecting volatile organic analytes directly in samples.
- Nanocantilever arrays detect and identify bacteria accurately at low concentrations in complex samples.

Further Reading

Capelli, L., Sironi, S., and Del Rosso, R. (2014). 'Electronic noses for environmental monitoring applications'. Sensors 14:11 19979–20007.
This review exemplifies the use of electronic noses.

Gui, Q., et al. (2017). 'The application of whole cell-based biosensors for use in environmental analysis and in medical diagnostics'. Sensors 17: 1623.
This paper reviews the use of whole-cell-based sensors.

Niemz, A., Ferguson, T. M., and Boyle D. S. (2011). 'Point-of-care nucleic acid testing for infectious diseases'. Trends in Biotechnology 29:5 240–50.
This paper provides an overview of nucleic acid-based point of care tests.

Ravikumar S., et al. (2017). 'Engineered microbial biosensors based on bacterial two-component systems as synthetic biotechnology platforms in bioremediation and biorefinery'. Microb Cell Fact 16: 62.
This paper reviews biosensors in the context of bioremediation.

Discussion Questions

7.1 Discuss the issues arising from low specificity and low sensitivity, and how to compromise if necessary.

7.2 Discuss point of care tests in the context of limited healthcare resources.

7.3 Discuss the benefits of using sensors remotely.

7.4 Discuss the principles of selecting a reporter gene.

8 MICROBIAL BIOTECHNOLOGY AND AGRICULTURE

Learning Objectives

- To learn that microbes will play an important role in the future of sustainable agriculture;
- to appreciate that the soil is a complex environment;
- to learn that microbial inoculants can be single strains or a consortium consisting of different strains of microbe;
- to appreciate that the challenges around the future of microbial inoculants are scaling from the laboratory to the field;
- to appreciate that the success of biological control agents has been limited by their ability to control a narrow range of pests, slow action, and short field life.

This chapter will consider the importance of microbial biotechnology to agriculture. This is crucially important as we have a growing population. The demands on agricultural production will increase 70 percent by the year 2050, by which time we will have to feed an estimated 10 billion people. Against a background of climate change and emerging plant pathogens, whilst also trying to preserve our planet and live sustainably, this is a big challenge. We will cover a range of topics in this chapter, including microbes as biological control agents and the use of bacteria as genetic tools and microbial inoculants. First, though we should consider the soil environment and crucially the issue of soil fertility.

8.1 The soil environment

The soil is a highly complex environment and soil composition is dependent on the geology of the land underneath the soil and the climate conditions. Table 8.1 provides further information on soil types and their characteristics.

Microbial populations in different soils will change depending on the organic and water content.

Soil fertility

Soil fertility is a complex issue. The soil has to contain the right nutrients at an appropriate level to support plant growth. There are at least 16 essential elements needed for plant growth including carbon, hydrogen, oxygen, nitrogen, and phosphorous. In addition to these essential elements the soil needs the right level of organic matter, has to be well draining, but also be able to retain moisture. Soils can be improved by the addition of fertilizers, but this increases the cost of food production. There are also other methods such as crop rotation which will increase soil fertility. Figure 8.1 shows a model of soil fertility management.

The degree of soil aggregation controls properties of the soil such as water availability, oxygen tension, and nutrient availability. The microbial component also has significant effect on the physical, chemical, and biological characteristics of the soil. Clay particles have the highest surface area for interactions with water molecules; they can also adsorb organic material and provide a surface for soil microbes to colonize.

Humic material is a rich layer of organic matter (Figure 8.2) is made up of undefined polymers and simple organic compounds. It is formed from the partial decomposition of plants, animals, and microbes. Humic material can bind small organic molecules, and this makes them unavailable for use by microbes. It can also cause soil minerals to aggregate into particles, termed microaggregates, which are <50 μm in size and contribute to the soil structure. Macroaggregates are larger than >50 μm and are stabilized by polysaccharides and microorganisms.

Table 8.1 Types of soil and their characteristics

Soil type	Water retention	Nutrient levels
Sandy	Fine grain, retains very little water, large air spaces, can be acidic	Poor level of nutrients
Clay	Little air space, poor drainage, retains water	High level of nutrients, trapped onto the clay particles
Peaty	High water content but is acidic	Rich in organic matter and high level of nutrients
Chalky	Alkaline soil, retains little water. Contains calcium carbonate or lime	High level of nutrients, but these might not always be available
Silty	Intermediate particle size. Can retain water and drain poorly	Medium level of nutrients
Loam	A perfect soil mix of 40% sand, 40% silt, 20% clay, fertile and well draining	Good nutrient levels

Figure 8.1 A stylized model of soil fertility management.

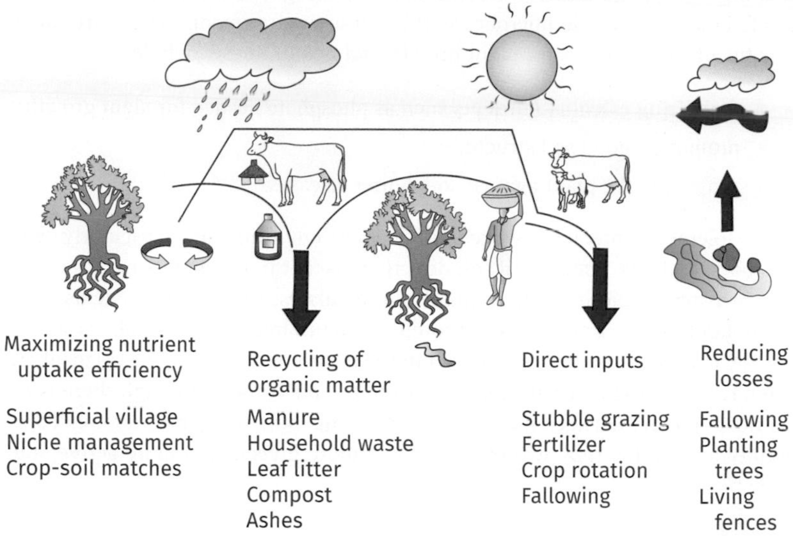

Figure 8.2 Humic material.

zlikovec/Shutterstock.com

Characteristics of the soil microbial population

Microbes such as bacteria, fungi, protists, and archaea constitute only a fraction of the soil mass (0.5 percent), but they have a major impact on the properties of the soil. The microbial biomass consists of growing cells to resting cells, but any growth of microbial cells in the soil is quite slow. Within different soil types, such as those described in Table 8.1, the population of microbes will vary in both their number and their diversity, for example fungi prefer to live in soil of a lower pH, and bacteria prefer soils of a higher pH. Typical cell numbers are

10^8 to 10^{10} cell/g soil, and the number of species from a few up to 10,000 species/g. This may still be an underrepresentation as we begin to know more about the viable but **unculturable** microbes within these environments. There are many benefits of soil microbes for plant growth and health, these include:

- mobilizing essential nutrients such as phosphate needed for plant growth;
- promoting good soil structure;
- protecting the plant against soil-borne diseases.

Many microbial interactions occur in the soil environment, particularly with the roots of plants. The area of soil directly adjacent to the plants roots is called the **rhizosphere** and it is rich in plant root exudates, such as amino acids, sugars, organic acids, carbohydrates, proteins, and vitamins. These exudates stimulate the microbial community and aid in the health of the plant. The rhizosphere extends to approximately 1 mm away from the plant root, although there is no district boundary between it and the surrounding 'bulk soil'. The microbial community around the rhizosphere tends to be more diverse and more active, and

Case study 8.1
The rhizosphere effect

The microbial population in cfu/g in the rhizosphere (R) of wheat and bulk soil (S) is shown in Table 8.2. The R/S ratio is the ratio between the number of microorganisms in the rhizosphere compared with the bulk soil.

Which groups of microorganism have the highest R/S ratio? Ammonifiers and denitrifers. Why do you think that this is the case? These microorganisms are essential in the nitrogen cycle, the ammonifiers take part in the process of ammonification (which is the release of ammonia from the decomposition of organic material) and the denitrifiers take nitrate and reduce it back to dinitrogen (see Figure 8.4).

Table 8.2 Microorganism population in the rhizosphere (R) of wheat and bulk soil (S) and an R/S ratio

Microorganism	Rhizosphere soil cfu g^{-1}	Bulk soil cfu g^{-1}	R/S ratio
Bacteria	1.2×10^9	5.3×10^7	23
Actinomycetes	4.6×10^7	7.0×10^6	6.6
Fungi	1.2×10^6	1.0×10^5	12
Protozoa	2.4×10^3	1.0×10^3	2.4
Algae	5.0×10^3	2.7×10^4	0.19
Ammonifiers	5.0×10^8	4.0×10^6	125
Denitrifiers	1.26×10^8	1.0×10^5	1260

microbes are present in larger numbers than in the bulk soil. This is called the 'rhizosphere effect' and it can be measured using the ratio of population density of the rhizosphere (R) compared with the population density of the bulk soil (S). The rhizosphere effect is explored further in the Case study 8.1. All of the important plant microbe interactions which we cover in later sections of this chapter occur in the rhizosphere.

There are some microbes, called endophytes, which are able to penetrate the plant cell tissues and live inside the plant in locations such as the roots, leaves, and fruits, without causing any disease symptoms. Many endophytes cannot be cultured, thus we are only aware of them through metagenomics. Soils, which have been damaged through natural or anthropogenic events can be bought back to health using bioremediation; this was explored in Chapter 5.

Key points

- The soil is a highly complex environment, with many different soil types.
- To maintain plant growth soil fertility has to be managed.
- The soil microbial community is vital for plant heath and growth.
- The rhizosphere is a crucial area of the soil where microbial activity is high and beneficial symbiotic relationships start to form.

8.2 Microbial inoculants

Microbial inoculants are either single microbial strains or microbe consortia (consisting of several different species) which are added either to the soil, or are coated onto seeds. They are reported to have a variety of beneficial effects, from solubilizing minerals such as phosphorus and potassium to increasing growth of the plant and product yield. Inoculants can also protect the plant against soil microbial pathogens. A range of commercial microbial inoculant products are provided in Table 8.3, along with their advertised benefits. There is a growing global market for microbial inoculants and by 2020 the market is estimated to be worth 1,295 million USD. Rhizobium inoculants have the biggest share of the market at 79 percent, followed by mycorrhizal inoculant products (15 percent) and phosphate mobilizing products (7 percent). China and India are the biggest producers and consumers of these products as there are tax incentives and grants to support their manufacture. However, it needs to be noted that these products have little regulation, and their effectiveness can be variable.

There are a number of steps to the development of a commercial inoculant product: the initial research phase, industry testing and formulation, followed by their use commercially in the field. The steps in these phases are described in Figure 8.3. The growers' requirements have to be taken into consideration as the use of inoculants must be compatible with their existing practices. In addition inoculants have to be:

- easy to use;
- compatible with seedling equipment;
- suitable in their storage properties, including having a long shelf life;
- able to be used in different field conditions, including different soil types.

Table 8.3 Examples of commercial microbial inoculant products

Product	Contains	Manufacturer	Benefits
Mycormax	Arbuscular mycorrhizal fungi *Glomus* spp.	JH Biotech Inc.	Physical barrier against root pathogens and nematodes Increased absorption of essential nutrients.
Diehard Ecto Root Dip	Ectomycorrhizal fungi *Trichoderma* fungi, *Pseudomonas* and *Streptomyces* bacteria	Horticultural Alliance	Boosts survival of trees. Quick root establishment. Improves nutrient availability
LegumeFix	Nitrogen-fixing bacteria *Bradyrhizobium* spp.	Legume Technology	Increased yield through nitrogen fixation
Soyflo	*Rhizobium* spp.	Soygro	Increased yield through nitrogen fixation
Bactofil	*Bacillus* spp.		Stimulates root growth and plant growth. Solubilizes non-soluble phosphorus and potassium

This section will now look more closely at these three major types of microbial inoculant:

1. Rhizobium inoculants;
2. Mycorrhizal inoculants;
3. Phosphate-mobilizing inoculants.

Rhizobium inoculants

Rhizobium inoculants have the biggest share of the soil inoculants market (79 percent). They are important in helping to increase the nitrate content of the soil. Nitrogen is a component of a variety of macromolecules, including amino acids, proteins, and nucleic acids. It is an essential nutrient and the lack of it will limit crop yield. Whilst nitrogen is the main component of the atmosphere (78 percent), the majority of organisms cannot directly assimilate gaseous nitrogen. To be used it has to be in a 'fixed' form, such as ammonia or nitrate. The only organisms which can fix gaseous nitrogen are bacteria. Nitrate fertilizers can be added to crops, but they are expensive and therefore add to the cost of production. In addition, they can cause nitrates to leach from the field and enter the water system. Excess nitrate causes environmental problems and can lead to eutrophication in streams and rivers.

Biological nitrogen fixation

In order to understand the importance of rhizobium inoculants, we need to examine the nitrogen geochemical cycle. First, we will look at the nitrogen cycle at the point of biological nitrogen fixation (BNF), which is the conversion of atmospheric nitrogen (N_2) to ammonium compounds, highlighted in Figure 8.4 in green.

Nitrogen fixation can occur due to lightning, which produces NO_x, but this fixes nitrogen in much smaller amounts than biological nitrogen fixation at

Figure 8.3 Flow diagram for development of bacterial inoculants. The steps in the development of a new product are split into three phases: the research phase, the industry phase, and finally commercialization.

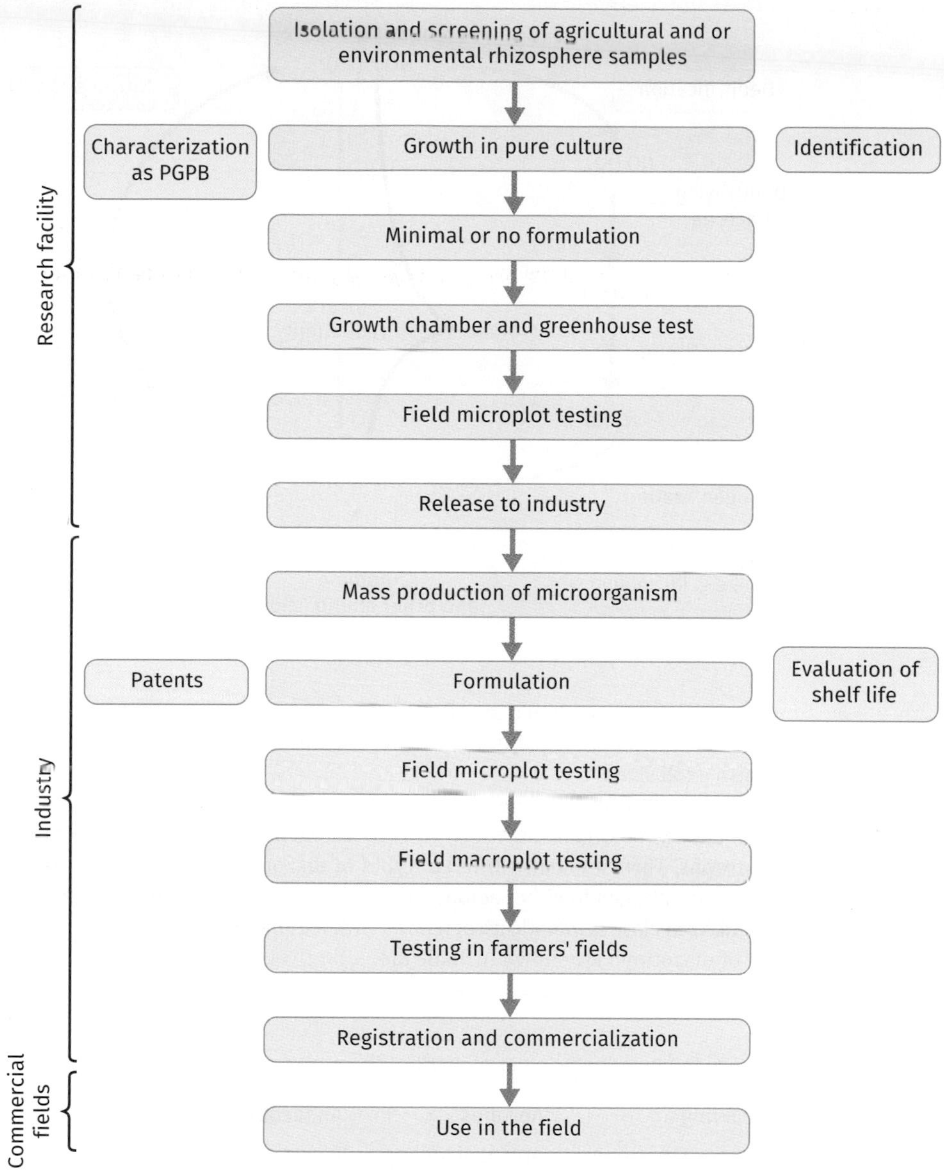

Figure 8.4 The nitrogen cycle.

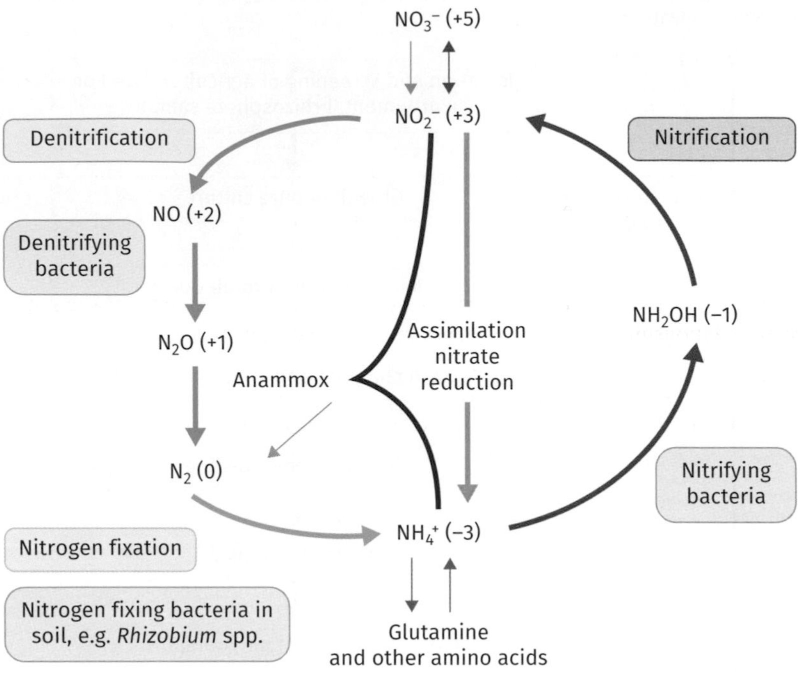

only 5 Tg N yr $^{-1}$. Ammonia can also be man-made through the Haber-Bosch process. Ammonia produced in this way is used in the production of fertilizers and also explosives.

Biological nitrogen fixation is only performed by prokaryotes, there are no known eukaryotic organisms which can fix nitrogen. BNF organisms are called **diazotrophs**. There are many different types of diazotroph; some are free living and can fix nitrogen in either aerobic or anaerobic conditions, some are symbiotic and form close symbiotic associations with leguminous plants. The different types of diazotroph are shown in Table 8.4.

Table 8.4 Examples of the types of diazotroph (nitrogen-fixing bacteria)

Free living	Aerobes	Anaerobes
Chemoorganotrophs	*Azotobacter* spp.	*Clostridium* spp.
Phototrophs	*Cyanobacteria*	
Chemolithotrophs	*Thiobacillus* spp.	
Symbiotic	**Bacteria**	**Plant**
Chemoorganotrophs	*Rhizobium* spp.	Peas, beans, clover
	Bradyrhizobium spp.	Soya
	Frankia spp. (Actinomycete)	Alder
	Sinorhizobium spp.	Alfalfa

All diazotrophs use the same enzyme to fix nitrogen, this is **nitrogenase**. It requires a lot of energy to drive the reaction as N_2 is very stable due to its triple bond; one N_2 molecule requires 16 ATP molecules to break. This is shown in the following equation:

$$N_2 + 8H^+ + 8e^- + 16ATP \rightarrow 2NH_3 + H_2 + 16\ ADP + 16P_i$$

The nitrogenase enzyme has an Fe-S-Mo centre involved in electron transfer. The nitrogenase enzyme is inactivated by the presence of oxygen, which means that the reaction is anaerobic.

Rhizobium–legume symbiosis

Rhizobium inoculants are produced from bacteria which form a mutualistic symbiotic relationship with plants; they often produce root nodules as part of the symbiosis. It is a mutualistic symbiotic relationship as both partners benefit. The bacteria get fixed carbon from the plant through photosynthesis and the plant gets ammonium compounds from the bacteria through BNF. The relationship can be quite specific, so the bacterium *Rhizobium* spp. will nodule peas, beans, vetch, and clover, *Bradyrhizobium* spp. will nodule soya plants, and *Sinorhizobium* spp. will nodule alfalfa. More examples are shown in Table 8.5.

It is important to understand why these microbes are so important to BNF before we examine their production as soil inoculants.

Steps in the development of the symbiosis

Symbiotic nitrogen-fixing bacteria such as *Rhizobium* spp. are often present in the rhizosphere, the thin layer of soil directly adjacent to the plant roots. However, as we will learn in the following section they can be deliberately added to the soil through the use of *Rhizobium* inoculants. The steps in the *Rhizobium*–Legume symbiosis are shown in Figure 8.5.

The development of the symbiotic relationship starts when the legume plant produces chemical signals called flavonoids in low nitrogen conditions. These flavonoids induce the transcription of a series of genes in the rhizobia bacteria called *nod* genes. Flavonoids can be very specific and only induce *nod* genes in compatible rhizobia. For example, *Rhizobium leguminosarium* will only induce its *nod* genes in the presence of specific flavonoids produced by plants such as peas and vetch. The *nod* genes can be found clustered together on a giant plasmid, called the Sym Plasmid, inside the bacterial cell. The *nod* genes code for proteins which synthesize nod factors.

Table 8.5 *Rhizobium* spp. and their host plants

***Rhizobium* spp.**	**Plant host**
R. trifoli	Clover (*Trifolium*)
R. meliloti	Alfalfa (*Medicago*)
R. phaseoli	Beans (*Phaseolus*)
R. leguminosarum	Pea (*Pisum*)
R. lupini	Lentil (*Lens*)
Bradyrhizobium japonicum	Soyabean (*Orthinopus*)

Figure 8.5 Steps involved in the development of root nodules on a leguminous plant.

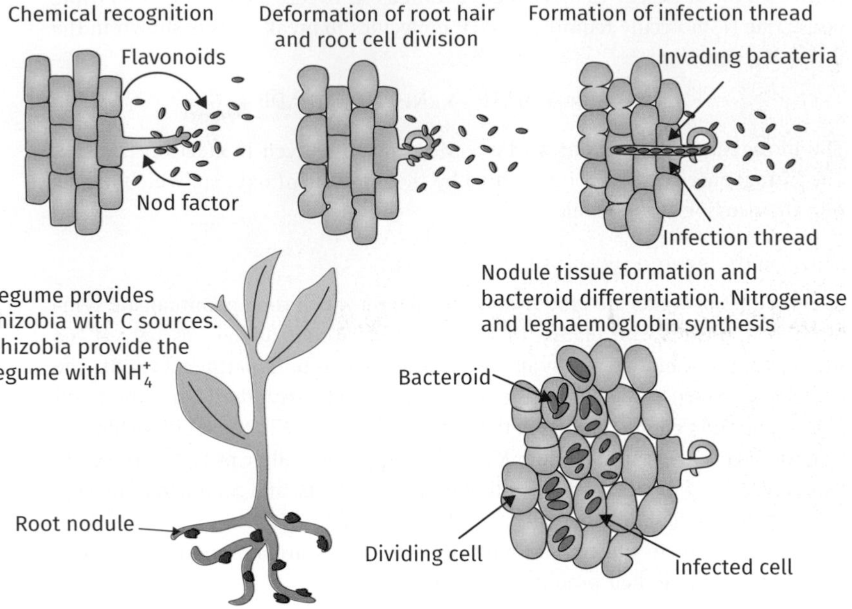

Nod factors are lipo-chitooligosaccharide signalling molecules which cause a root hair of the plant to deform and start to curl. This action traps the rhizobia, which can then form an infection thread which runs through the root cortical cells of the plant host. The Nod signalling factors also cause the root cortical cells to swell and divide, and this produces the root nodule itself. When inside the cytoplasm of the root cortical cells, the rhizobia differentiate into structures called **bacteroids**. In most cases this is a permanent differentiation process; the bacteroids can never return to being free-living bacteria. These bacteroids then carry out biological nitrogen fixation through the induction of another cluster of genes called *nif* genes, which code for the nitrogenase enzyme and the *fix* genes which are needed for the nitrogen fixation process. The regulation of nitrogenase is tight and the expression of *nif* structural genes are controlled by *nifA*. This is a positive regulator, the presence of *nifA*, allows the expression of the *nif* genes.

A compound called leghaemoglobin, which is produced by the plant, protects the nitrogenase enzyme which is susceptible to oxidative damage within the root nodule. Leghaemoglobin is almost identical to our own haemoglobin molecule and is responsible for the pinky tinge of the root nodules (shown in Figure 8.6). It is the most abundant protein in the root nodule. The iron in the leghaemoglobin binds and transports the oxygen around the plant. Nitrogen fixation requires a lot of energy, so the leghaemoglobin delivers the oxygen to the mitochondria in the plant cells and also to the bacteroids, but it also protects the nitrogenase enzyme from oxidative damage.

Figure 8.6 Root nodules of *Rhizobium leguminosarum* clearly showing the pink tinge of leghaemoglobin.

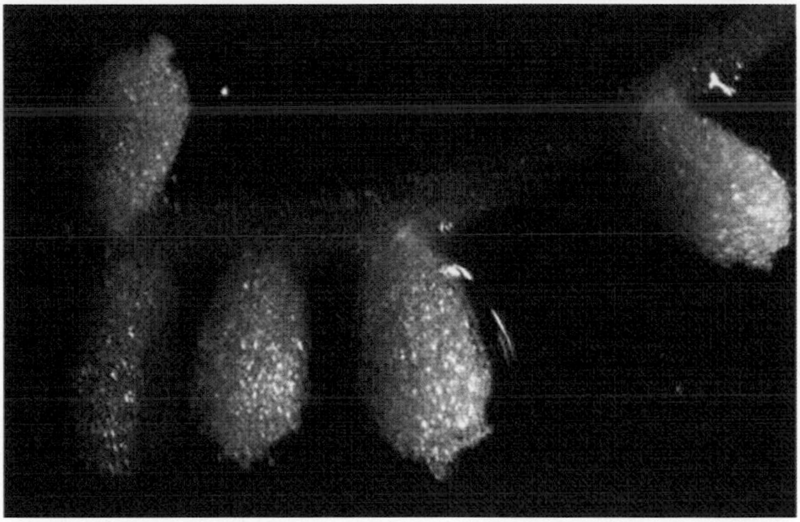

Dr. Jeremy Burgess/Science Photo Library

Characteristics and production of *Rhizobium* inoculants

The different species of *Rhizobia* are formulated into different microbial inoculant products (Table 8.5).

There are specific attributes which the inoculants need to have in order for them to be successful.

- **effective**—the inoculant must be able to have a symbiosis with the right plant;
- **competitive**—the inoculant must be able to survive in the soil environment and stimulate nodule development on the roots;
- **stress tolerant**—the inoculant must be able to stimulate nodule formation over a range of soil temperature, pH, and salinity;
- **persistent**—the inoculant must be able to maintain its numbers in the soil;
- **shelf life**—the inoculant must be able to maintain its viability during storage before use.

To make the inoculant, the specific *Rhizobium* species is grown in shake flask culture, which is then used to inoculate a fermenter. The growth medium contains complex sources of carbon and nitrogen, for example tryptone and yeast extract. The length of the fermentation will depend on which species of *Rhizobium* is being grown; *R. leguminosarum* takes three days to reach the correct cell density, whilst *B. japonicum* can take five.

After fermentation, the *Rhizobium* culture is mixed with the carrier in a ratio of approximately 1 litre of culture to 1 kg of carrier to have a bacterial count of 10^9 bacterial cells per g.

Carriers should be non-toxic, have a neutral pH, have a high water-holding capacity, be able to protect the bacteria from harmful factors such as sunlight, be sterilized, and be inexpensive. There are three main types of carrier for the bacteria:

1. **Peat**—this is the oldest and most common carrier, the bacteria are added to sterilized peat. A disadvantage is that peat is not available in all countries.
2. **Granular**—this is made up from peat, clay, charcoal, sawdust, or compost.
3. **Liquid**—the bacteria are suspended in the liquid, e.g. carboxymethyl cellulose, glycerol, and water, and the liquid is poured into a furrow or put directly onto the seed.

Figure 8.7 Formulations of inoculants for agriculture and environmental uses.

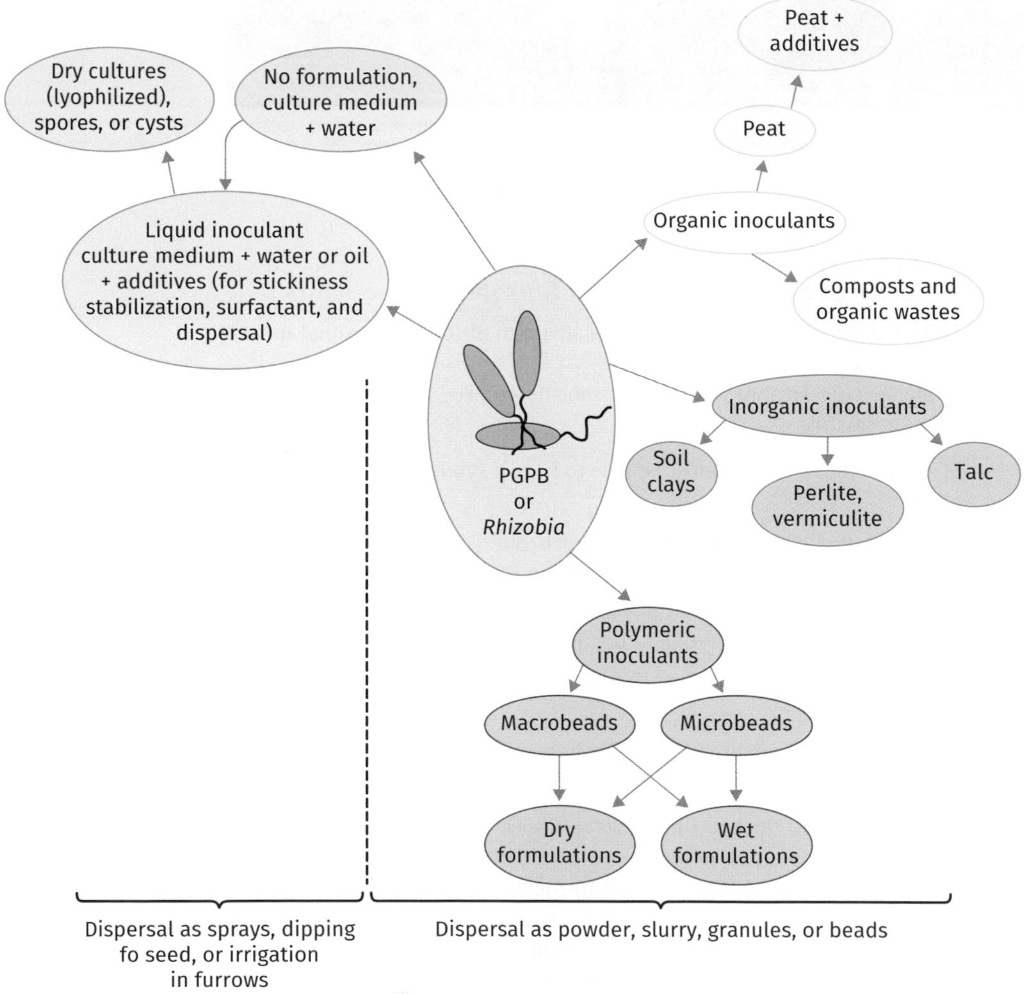

Adapted by permission from Springer Nature. Plant and Soil, Advances in plant growth-promoting bacterial inoculant technology: formulations and practical perspectives (1998–2013), Yoav Bashan, Luz E. de-Bashan, S. R. Prabhu et al, Copyright © 2013, Springer Science Business Media Dordrecht

After the carrier and bacteria have been mixed, it is cured for 14 days at 25–30°C to allow for further bacterial growth. It is then packaged and shipped to farmers. When the farmers receive their inoculant housed in its carrier, it will needed sticking to their seed. There are several way to do this; one of the most common is the use of 40 percent (w/v) gum arabic mixed in hot water. Other 'stickers' include corn syrup, honey, powdered milk, and vegetable oils. There are circumstances where sticking the inoculant to the seed will not work, for example when the seeds have been treated with pesticides which could skill the rhizobia. In these instances the inoculant is sprayed directly onto the soil.

Famers are advised that if they are in doubt about the effectiveness of their inoculant, they should conduct a 'grow out test'. This is where the inoculant would be added to the seeds and then planted in pots under controlled conditions. Controls would also have to be established which have not been treated with inoculant. Growth would be observed and the plants dug up when 50 percent are flowering. The root nodules are then counted per plant and cut open to look at their colour which should be pink (Figure 8.6).

The major producers of *Rhizobium* inoculants are Australia, Brazil, New Zealand, USA, and Thailand. Table 8.6 highlights some commercially available products.

The market for microbial inoculants is set to grow, with more pressure on the world's ability to produce enough food for the population. Other products are being developed and one such area involves cyanobacterial inoculants, the Bigger picture 8.1 explores this in more detail.

Mycorrhizal inoculants

Mycorrhizal inoculants represent the second largest global market for inoculants (15 percent). Mycorrhizae are mutualistic symbiotic associations that form between the roots of most species of plants and particular fungi. The relationship develops and is maintained within the soil environment. There is bidirectional movement of nutrients where fixed carbon from photosynthesis by the plant flows to the fungus and inorganic nutrients from the soil move to the plant. Mycorrhizal fungi form a crucial link between the plant root and the soil.

As much as 20 percent of the total carbon fixed by the plant through photosynthesis is transferred to the fungus and the plant increases photosynthetic activity to compensate for the loss. The flow of carbon to the soil mediated by mycorrhizal fungi serves several important functions:

- fungal hyphae produce hydrolytic enzymes which increase nutrient availability by breaking down complex organic compounds;
- fungal hyphae bind soil particles together, thereby improving soil aggregation and characteristics such as drainage properties;
- a unique rhizosphere microbial community develops, called the mycorrhizosphere, which is the area of soil adjacent to the fungal hyphae.

Table 8.6 Examples of commercial *Rhizobium* inoculant products

Product	Manufacturer	Species	Formulation	Plant
SoyRhizo®	XiteBio	*Bradyrhizobium japonicum*	Liquid	Soyabean
PeasRhizo®	XiteBio	*Rhizobium leguminosarum*	Liquid	Peas and lentils
Nitragin™ Gold	Monsanto BioAg	*Rhizobium meliloti* and *R. leguminosarum* biovar *trifoli*	Granular	Alfalfa, melilot, clover

The bigger picture panel 8.1
Cyanobacterial inoculants

Rice is important as a staple food crop for over half of the world's population, and some 143 million ha of low lying land is flooded for use in rice cultivation. Cyanobacteria are essential for the long term stability of rice paddy fields. Cyanobacteria are phototrophs, meaning that they can photosynthesize. They are also diazotrophs and carry out nitrogen fixation at night when they are not producing O_2 through photosynthesis.

Multicellular cyanobacteria, for example *Nostoc* sp. in grasslands and *Anabena* sp. in rice paddies, are filamentous and have specialized cells for N_2 fixation called **heterocysts**, which can be seen as light brown structures in Figure 8.8. Heterocysts are thick-walled cells which occur after every 10–15 cells. These heterocysts do not contain photosynthetic systems and have specialized membranes with low O_2 levels and high N_2 diffusion.

Rice farmers can add Cyanobacteria to their paddy fields. To do this, starter cultures are purchased which can then be grown in large enough quantities to act as fertilizer. However the scale-up of growth is reliant on the skill of the farmer, there appear to be no commercial fertilizers manufactured currently.

Figure 8.8 Cyanobacterium with heterocysts.

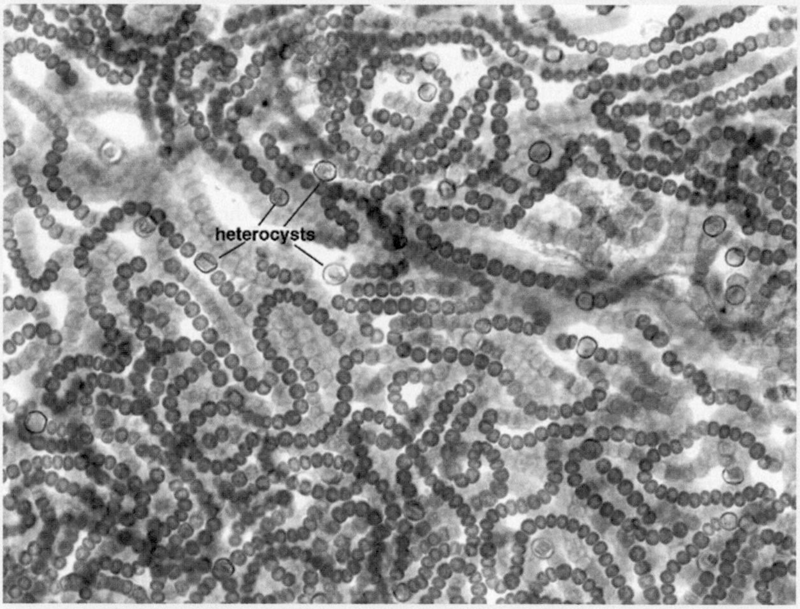

Dr. David Dalton

There are two main types of mycorrhizal association. Ectomycorrhizal (EcM) fungi surround the root cortical cell in a sheath-like net called a **Hartig net**. These types of association are found on woody shrubs and forest trees, and their fruiting reproductive structures can often be found on the woodland floor. A good example is the fungus *Amanita muscaria* (Figure 8.9(a) and (b)) which forms an ectomycorrhizal association with silver birch trees.

Endomycorrhizal fungi, also known as arbuscular mycorrhizae (AM), develop a highly branched structure called an arbuscule inside the root cortical cell. The fungi that form arbuscular mycorrhizae are all classified in the fungal phylum, Glomeromycota.

Challenges for mycorrhizal inoculants

There are several challenges associated with producing and using mycorrhizal inoculants. There is difficulty in culturing the obligate AM on a large scale in the absence of the host plant, but, despite this, companies are producing inoculants

Figure 8.9 (a) **Fruiting body of Amanita muscaria in close proximity to silver birch.** The hyphae of the fungus are in association with the roots of the tree in an ectomycorrhizal relationship; (b) the Hartig net is shown surrounding the root cortical cells.

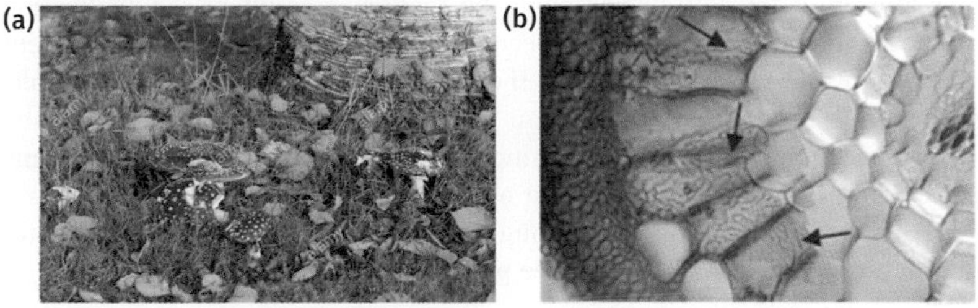

(a) MichaelGrant/Alamy Stock Photo; (b) Mark Brundrett

Figure 8.10 An arbuscule shown inside the root cortical cell. A is the arbuscule and C is the fungal hyphae.

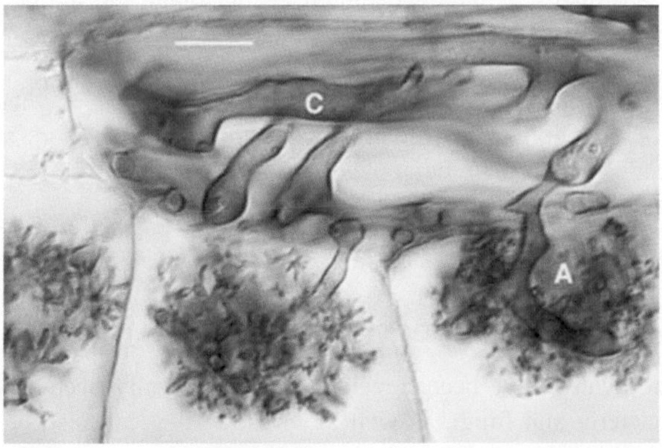

Mark Brundrett

(Table 8.3) which are used in agriculture, forestry, and horticulture. Formulations include the hyphae, spores, and also root fragments of host plants—these are added to suitable carriers.

Inoculating with 'native' species tends to be better than using 'foreign' inoculants, many commercially available inoculants, when tested, fail to establish a symbiosis. There are many local factors which need to be considered such as soil moisture, pH, and nutrient levels. There is also the issue of the disruption of the local soil community which can have a detrimental effect.

Phosphate-mobilizing inoculants

After nitrogen, phosphorus is the most limiting nutrient for plant growth. The element is quite abundant (0.7 percent of the Earth's crust) but the underlying geology of the soil will determine how much phosphorus is present. Its bioavailability regardless of its concentration in the soil is dependent on factors such as pH, moisture, aeration, temperature, and organic matter content. Phosphorous can be added to the soil as an inorganic fertilizer. However, this is not sustainable as there is a limited reserve of rock phosphate which can be mined for making phosphate fertilizer. Most of the available rock phosphate is in Africa.

Strategies to lessen our dependence on rock phosphate include the recycling of phosphorous from human and animal waste and releasing the phosphorous already in the soil, (which could be there as a result of the application of phosphate fertilizer) which has become unavailable. A number of different microbes in the soil can help with the release of phosphorous, for example, species from the *Rhizobia* genera can solubilize phosphorus. They do this by secreting low molecular weight organic acids which can then act on inorganic phosphorus, e.g. 2-ketogluconic acid and also acid phosphatases. Mycorrhizal fungi can translocate phosphate through their hyphae. In addition, some bacteria can both solubilize and mineralize phosphorous.

Key points

- Microbial inoculants are either single microbial strains or consortia.
- Microbial inoculants benefit plant growth and health by mobilizing essential minerals and protecting against pathogens.
- There are many different commercial microbial inoculants, but the *Rhizobium* inoculants, which promote biological nitrogen fixation, have the biggest share of the market.
- Successful microbial inoculants have specific characteristics. They must be effective, competitive, stress tolerant, persistent, and have a long shelf-life.
- Microbial inoculants can be applied to the seed or to the soil in which the seed is planted.

8.3 Microbial biopesticides

Microbial pesticides have been developed from naturally occurring microbes, including bacteria and fungi, as well as viruses. They can also be considered to be microbial inoculants. Biological control agents are examples of microbial pesticides and they can be defined as the use of one organism to control another.

They are an attractive alternative to chemical pesticides as they have a number of benefits:

- safer than chemical pesticides with a smaller risk to human health;
- less damaging to the environment;
- targeted activity;
- effective in small quantities;
- decompose more quickly than chemical pesticides;
- incidence of resistance is reduced.

Microbes, mainly bacteria and fungi, have been used as biological control agents for nearly 40 years, but their use has failed to keep pace with new chemical pesticides. The success of biological control agents has been limited by their ability to control only a narrow range of pests, by their slow action, and their short field life. Table 8.7 provides examples of microbes which have been formulated as commercial biological control products.

In this section we will consider a number of examples of biological control agents in more detail, including *Bacillus thuringiensis* (Bt) and the mycopesticide Green Muscle™.

Bacillus thuringiensis (Bt)

Bacillus thuringiensis (Bt) is the most successful bioinsecticide and has been used commercially for over 40 years. It is sold to control a number of plant pests, including caterpillars, mosquito larvae, and black flies. Bt can be grown easily in a bioreactor using molasses as the carbon source. The products are sold as powders, which contain the spores of Bt, as well as the crystallized toxin.

Bacillus thuringiensis is a facultative anaerobe and lives in both soil and aquatic environments. The bacterium has inclusion bodies which are formed during sporulation (Figure 8.11). These are crystals of the Cry and Cyt proteins, produced from *cry* and *cyt* genes, respectively, which are plasmid encoded. The Cry and Cyt proteins are endotoxins. The α-endotoxins include the Cry proteins (they have α-helices)—the β-barrel are the Cyt proteins.

Table 8.7 Examples of microbes formulated as products of biological control

Product	Manufacturer	Biological control agent	Target disease/pest
Thuricide-Southern AG	Bonide	*Bacillus thuringiensis*	Caterpillars, cabbage worms
BlightBan™	Nufarm	*Pseudomonas fluorescens* A506	Frost damage, *Erwinia amylovora*
Green Muscle™	BCP and NPP	*Metarhizium anisopliae*	Locusts
EcoGuard™	LabanonTurf	*Bacillus licheniformis*	Dollar spot (a fungal disease)
Deny™	CCT	*Burholdaria cepacia*	*Rhizoctonia*, *Phythium*, and *Fusarium*
Nogall™	Bio Care Technology	*Agrobacterium radiobacter*	Crown gall disease
Mycostop™	Kemira OY	*Streptomyces griseoviridis*	Root infecting fungi, e.g. *Fusarium*
RootShield™	BioWorks	*Trichoderma harzianum*	Root-infecting fungi

Figure 8.11 (a) *Bacillus thuringiensis*; (b) cry toxin protein crystals.

(a) (b)

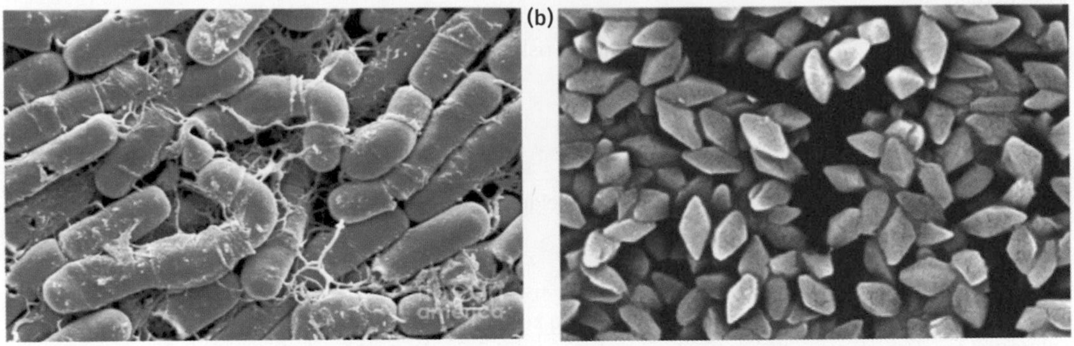

(a) Mediscan/Alamy Stock Photo; (b) P. R .Johnston/Wikimedia Commons/Public Domain

Look at Figure 8.12. The Lepidoptera digestive tract has three regions: pre-intestine, mid gut, and the hind gut. After the Bt toxin has been consumed (1), the crystals dissolve in the alkaline conditions of the mid gut (2). Protoxins are then released through the action of protease enzymes; this results in active toxins of 60–70 kDA in mass (3). The protoxins bind to specific receptors on the microvilli of the mid gut (4) and (5). The toxins are pore-forming, which means that pores are created in the membrane which lead to osmotic imbalance between the intracellular and extracellular environments. The microvilli are destroyed, the insect stops feeding and starves to death.

Figure 8.12 Mode of action of the Bt toxin inside the gut of the Lepidoptera larvae.

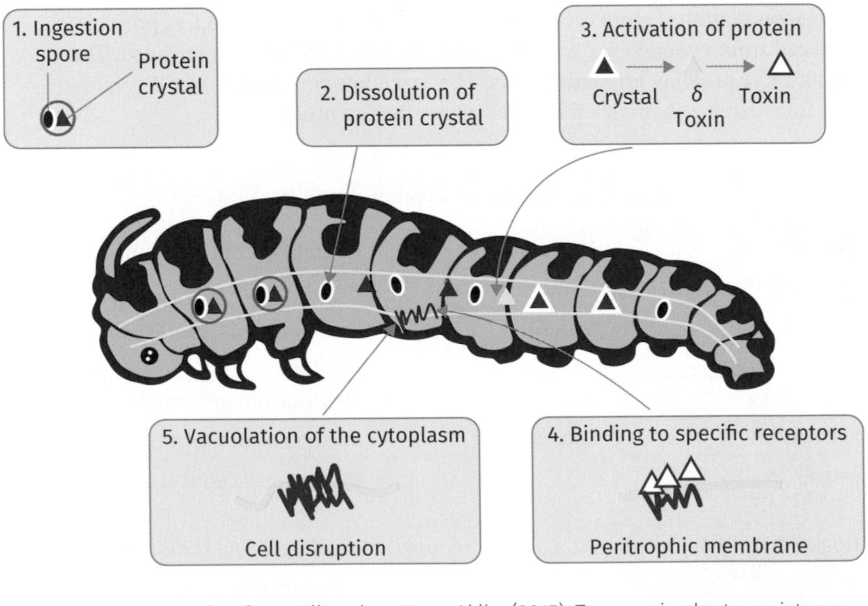

Adapted with permission from Wijerathna Yapa, Akila. (2017). Transgenic plants: resistance to abiotic and biotic stresses. Journal of Agriculture and Environment for International Development. 111. 245-275. 10.12895/jaeid.20171.643

There are issues with the use of the Bt toxin. It is not very persistent in the environment and the Bt toxin is sensitive to solar UV. It is important to note here that the Bt toxin gene has been cloned into a variety of plant species (see section 8.x).

Mycopesticides

A mycopesticide has been developed called Green Muscle® (Figure 8.13), which is based on the spores of the insect pathogen fungus *Metarhizium anisopliae*. Green Muscle® is available as a spore powder or an oil miscible concentrate, and is sprayed using normal equipment. The product is produced by two companies: BCP (Biological Control Products) of South Africa, and NPP (Natural Plant Protection) of France.

Mycoparasites are fungi which parasitize other fungi. They obtain all their nutrition from their living host; thus, they have the potential to be microbial pesticides. A good example is the control of *Heterobasidium annosum*.

Heterobasidium annosum is the causative agent of butt-rot, which damages timber at the base of the trunk. It is a serious pathogen of conifers in Britain, continental Europe, and North America. It spreads slowly by mycelial growth along the roots of diseased trees and infects healthy trees by root-to-root contact. This disease can be controlled by the hyphal interference of *Peniophora gigantea*. Spores are sold as suspensions in sachets, with the inclusion of a dye so that the forester ensures complete coverage of the stump (Figure 8.14).

Seedlings are often attacked by pathogens either pre- or post-germination. These diseases are collectively known as 'damping off' diseases or seedling blights. Glasshouse crops such as tomatoes are particularly susceptible, so a lot of research has been focused towards biological control agents for these glasshouse diseases. The fungus *Trichoderma harzianum* is a soil-dwelling filamentous fungus and is an effective biological control agent for damping off diseases in glasshouse crops. Commercial products of *T. harzianum* include Amino T made by AGRI nova Science and Bioharz™ liquid, made by International Panaacea Limited. Seeds are treated directly with these products, or in the case of Bioharz™ the seedling roots can be dipped into the liquid.

Figure 8.13 Mycopesticide products from *Metarhizium anisopliae*.

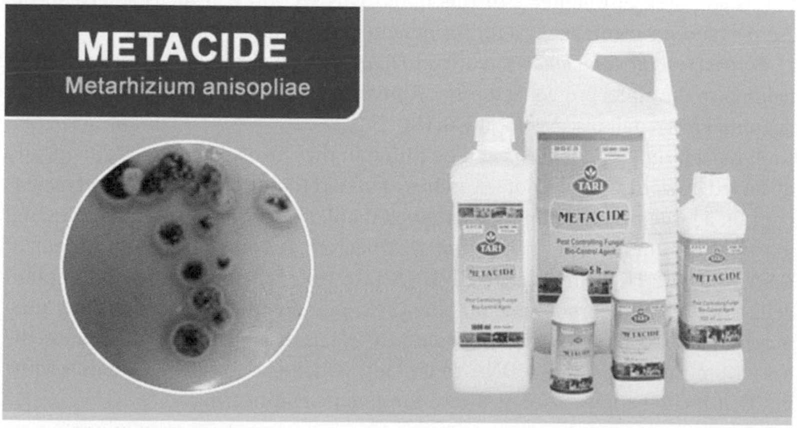

Figure 8.14 *Peniophora gigantea* spores covering a tree stump. The spore suspension has been coloured blue to ensure complete coverage of the stump.

Key points

- Microbial pesticides can be considered as microbial inoculants, they are used as an alternative to chemical pesticides.
- Biological control agents are examples of microbial pesticides.
- There are a number of commercially available biological control agents, the most widespread and successful being *Bacillus thuringiensis*.
- The Bt toxin gene has been cloned into a number of plant species.
- Fungi have also been developed as biological control agents.

8.4 Microbes as tools for genetic modification

Microbial enzymes have been crucial in the development of recombination technology, these have included the use of restriction enzymes for the digestion of DNA, as well as enzymes such as Taq polymerase for PCR. More recently the CRISPR Cas technology has been used for genetic modification; this system is explored in Chapter 1 of this book. This section will focus solely on the *Agrobacterium*-mediated system for genetic modification of plants.

Agrobacterium tumefaciens is a rod-shaped Gram-negative bacterium which belongs to the alpha proteobacterium. *Agrobacterium tumefaciens* is the causative agent of crown gall disease (Figure 8.15).

Agrobacterium tumefaciens can be found in the rhizosphere and is chemically attracted towards plant root exudates. Pathogenic strains of *A. tumefaciens* contain a large plasmid (approximately 200 kb), called the Ti (tumour-inducing) plasmid. It is quite remarkable that this bacterium can transfer a segment of DNA from the Ti plasmid, called T-DNA into the nucleus of the infected plant cell, where it is incorporated into the host genome and transcribed, this ultimately leads to the crown gall disease. It is fascinating not only because the bacterium transfers its own DNA to a plant, which is a different domain of life, but that it can be harnessed as a tool for genetic engineering.

Figure 8.15 (a) *Agrobacterium tumefaciens* attached to a plant cell; (b) Crown gall disease.

(a) (b)

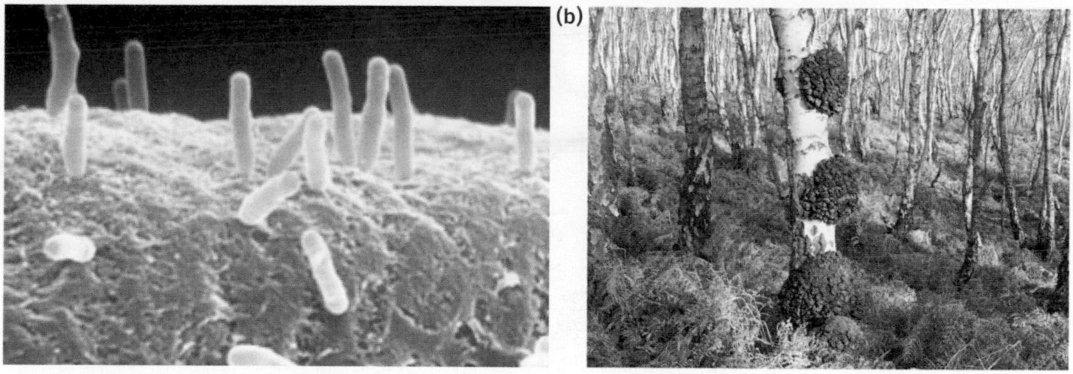

(a) Reproduced with permission from Martha Hawes (b) Caroline Morgan

The Ti plasmid encodes the genes needed for transfer into the plant cell. The T-DNA specifically encode the following genes:

- **Oncogenic genes**—encoding enzymes involved in the synthesis of auxins and cytokines which form the tumour.
- **Opine genes**—these are genes which encode proteins for the synthesis of opines. The opines are produced inside the tumour cells as a condensation reaction between amino acids and sugars. The opines act as both a carbon and nitrogen source for the growth of *A. tumefaciens*.

The transfer of the T-DNA is mediated by the 30–40 kb *vir* region of the Ti plasmid. The essential steps in the transfer of the T-DNA are shown in Figure 8.16. There are six essential operons in this *vir* region: *virA*, *virB*, *virC*, *virD*, *virE*, and *virG*; and two non-essential (*virF* and *virH*). *virA* is expressed constitutively, this gene is the sensor gene of a two-component regulator and it codes for a sensor protein (VirA) which can detect phenolic molecules released when plants are wounded [1]. This directs the bacteria to the wound site. When activated VirA phosphorylates the aspartate residue of the VirG protein; this is a DNA binding protein of the two-component regulator, and acts as a transcription factor, enabling the expression of the other *vir* genes [2].

The T-DNA is flanked both left and right by 25 bp highly homologous direct repeats. VirD1 and 2 are endonucleases, they cut the left and right side of the T-DNA segment on the lower DNA strand resulting in a single strand of T-DNA [3]. The VirD2 remains covalently bound to the T-DNA. The T-strand-VirD2 complex is translocated to the plant cell by a *virB*-encoded type IV secretion system (T4SS) [4]. The single stranded *virE2* gene is also transported to the plant cell using the same secretion system. The protein VirE2 is important, as when inside the plant cell it coats the T-strand to prevent nuclear degradation and also protects the 3' end prior to integration into the host genome [5]. The T-DNA is integrated into the host plant cell genome in random locations [6].

Figure 8.16 Mechanism of T-DNA transfer by *Agrobacterium tumefaciens*.

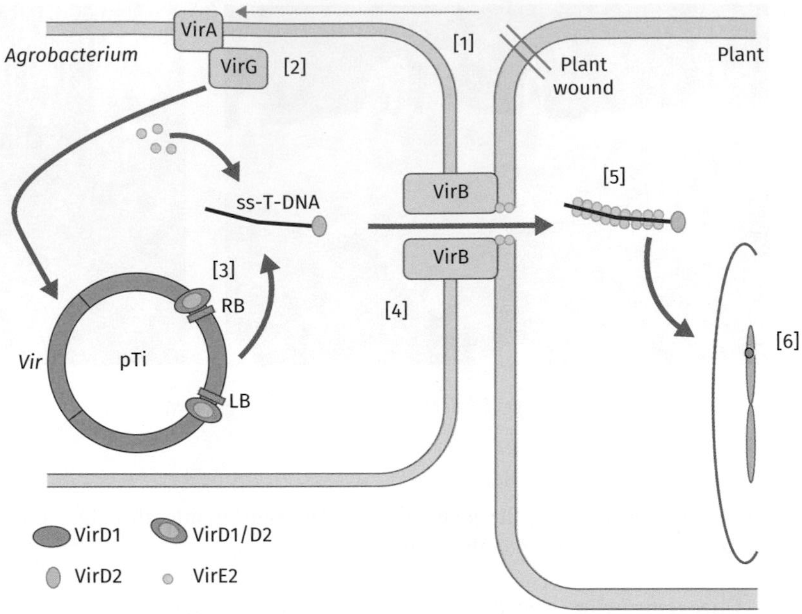

Use of *Agrobacterium* in genetic engineering

For genetic engineering purposes the Ti plasmid, which is very large (200 kb), may be divided into two. One plasmid encodes the *vir* operons required for the transfer (this is the disarmed Ti Plasmid or help plasmid)—see Figure 8.17(a). The other plasmid is the wide host range plant expression vector and this plasmid contains the T-DNA segment with its flanking ends, but the genes for opine production and the oncogenic genes have been removed—see Figure 8.17(b). The genetic manipulation of the plant expression vector is done in *E. coli*. There is a multiple cloning site (MCS) downstream of a plant promoter and a gene encoding resistance to an antibiotic, which can be used to select for transformants in the *E. coli* host and a herbicide resistance gene to a herbicide which the eventual host plant is susceptible to. The gene to be transferred to the plant is cloned into the MCS. This new construct is then transferred into an *Agrobacterium* strain using conjugation, which contains the 'disarmed' Ti Plasmid, which is also known as the helper Ti plasmid. This strain is then used to infect the plant.

Over 90 percent of the currently available commercial genetically modified crops are made using the *Agrobacterium* Ti system. Table 8.8 provides some examples.

Figure 8.17 The binary system for the transfer of cloned DNA into a host plant. (a) The help Ti plasmid; (b) wide host range vector for cloning the gene of interest.

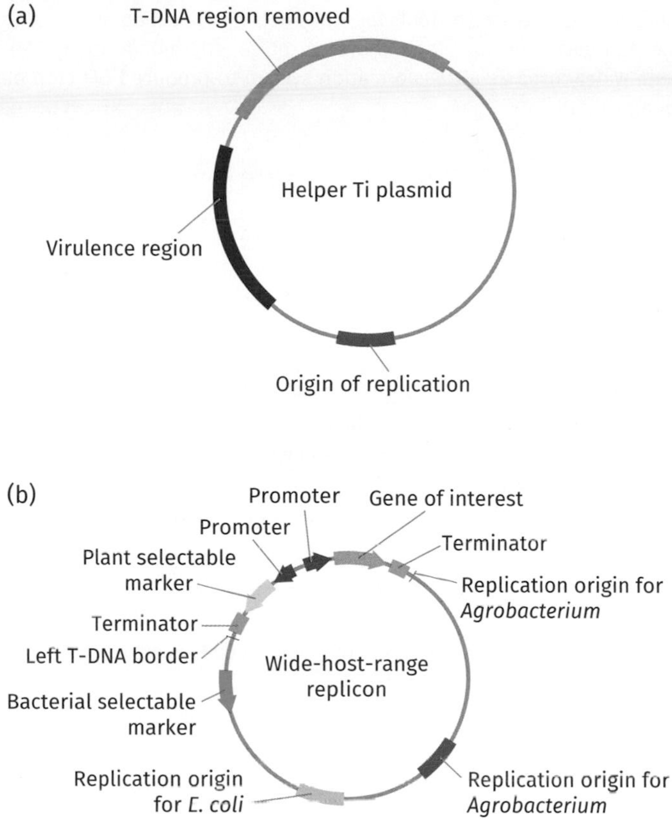

Table 8.8 Examples of commercial transgenic plants created using the *Agrobacterium* Ti system

Commercial transgenic plants			
Crop	**Name**	**Company**	**Properties**
Tomato (1994)	Flavr Savr	Calgene	Vine-ripened flavour and shelf life
Tomato (1994)		Zeneca	Consistency of tomato paste
Cotton (1996–1997)	Bollgard	Monsanto	*B. thuringiensis* toxin for insect resistance
Potato (1996–1997)	NewLeaf	Monsanto	*B. thuringiensis* toxin for insect resistance
Maize (1996–1997)	YieldGuard	Monsanto	*B. thuringiensis* toxin for insect resistance
Soybean Canola Cotton (1995–1996)	Roundup Ready	Monsanto	Glyphosate herbicide resistance

Key points

- Microbes produce enzymes such as restriction endonucleases and Taq polymerase which are used as tools for the genetic modification of crop plants.
- The *Agrobacterium tumefaciens* system of genetic modification (GM) has been widely used as a transformation system to produced GM crop plants for agriculture.

Chapter Summary

- Microbes will play an important role in the future of sustainable agriculture due to the increased demand on agricultural production.
- The soil is a complex environment, the characteristics of which are dependent on the underlying geology.
- A fertile soil needs to have the right level of nutrients to support plant growth.
- Microbes only represent a small amount of the actual mass of soil, but they have large impact on soil properties.
- The rhizosphere is a crucial area of the soil where microbial activity is high and beneficial symbiotic relationships start to form.
- Microbial inoculants can be single strains or a consortium consisting of different strains of microbe.
- Biological nitrogen fixation is carried out by prokaryotic bacteria which are known as diazotrophs.
- Rhizobium inoculants have the biggest share of the commercial market.
- Rhizobium inoculants must be effective, competitive, able to tolerate stress, persistent in the soil, and have a long shelf life.
- Challenges for the production of mycorrhizal inoculants include the culturing of the fungi on a large scale and the establishment of the mutualistic symbiosis in the field.
- Microbial biopesticides can also be considered to be microbial inoculants.
- The success of biological control agents has been limited by their ability to control a narrow range of pests, slow action, and short field life.
- Bt (*Bacillus thuringiensis*) is considered to be the most successful example of a biological control agent.
- Both fungi and bacteriophages have also been developed as biological control agents.
- Microbes can also be used as tools in genetic modification. A good example is the *Agrobacterium*-mediated system for genetic modification of plants.

Further Reading

Barea, J. M. (2015) 'Future challenges and perspectives for applying microbial biotechnology in sustainable agriculture based on a better understanding of plant-microbiome interactions'. Journal of Soil Science and Plant Nutrition 15(2): 261–82.

This article is a good, readable review of the problems, challenges, and opportunities for using microbes within agricultural systems.

From flask to field: How tiny microbes are revolutionizing big agriculture https://theconversation.com/from-flask-to-field-how-tiny-microbes-are-revolutionizing-big-agriculture-67041

The article below is a nice conversation piece on the use of phosphate mobilizing bacteria within the soil and the issues with scaling-up into the field.

Kuypers, M. M. M., Marchant, H. K., and Kartal, B. (2018). 'The microbial nitrogen-cycling network'. Nature Reviews Microbiology 16: 263–76. https://www-nature-com.ueaezproxy.uea.ac.uk:2443/articles/nrmicro.2018.9

If you would like to get to grips further with the impact which microbes have in nitrogen cycling, then this is an excellent review which takes the reader through the different transformation steps.

Yadav, A. N., and Saxena, A. K. (2018). 'Biodiversity and biotechnological applications of halophilic microbes for sustainable agriculture'. Journal of Applied Biology and Biotechnology 61(1): 48–55. http://www.jabonline.in/admin/php/uploads/255_pdf.pdf

The following is a journal article looking at the application of microbes which can tolerate high salt concentrations to aid plant growth. Salinity is an issue which affects large areas of cultivated land. This links to other aspects of this book (see Chapter 9) which look at extremophiles.

Discussion Questions

8.1 Microbial inoculants will never reach their full potential due to the issues of scaling from the laboratory to the field. Discuss.

8.2 What role will microbes play in the future of sustainable agriculture?

8.3 Why do we need to understand more about the microbial communities of different soils before we start to manipulate them?

9 USING EXTREMOPHILES IN BIOTECHNOLOGY

Learning Objectives

- To be able to give an overview of how microorganisms can survive and grow in extreme environments;
- to be able to explain how such extremophile adaptations can have applications in biotechnology;
- to be able to discuss examples of such extremophile applications in their context;
- to be able to discuss how such extremophile adaptations can be responsible for contaminating and impairing industrial processes.

Microorganisms (bacteria, archaea, fungi) can survive and grow in extreme environments with temperatures as high as 122° C and as low as −12° C, with pressure as high as 100 MPa, up to saturating levels of salinity and all pH values. For all these parameters and their combinations, individual organisms have a minimum and maximum level that supports growth, and an optimum, where growth is fastest. Outside these levels microorganisms are just about viable but cannot grow, and viability ceases further away from these levels. Life in such extreme habitats requires adaptations of biological molecules, processes, and structures which offer a multitude of applications in biotechnology (see Figure 9.1). We can use the organisms directly

Figure 9.1 The scope of classical biocatalysis (the square) is extended by extremozymes. Examples of applications are given for the contexts of extreme temperature, pH, and solvents.

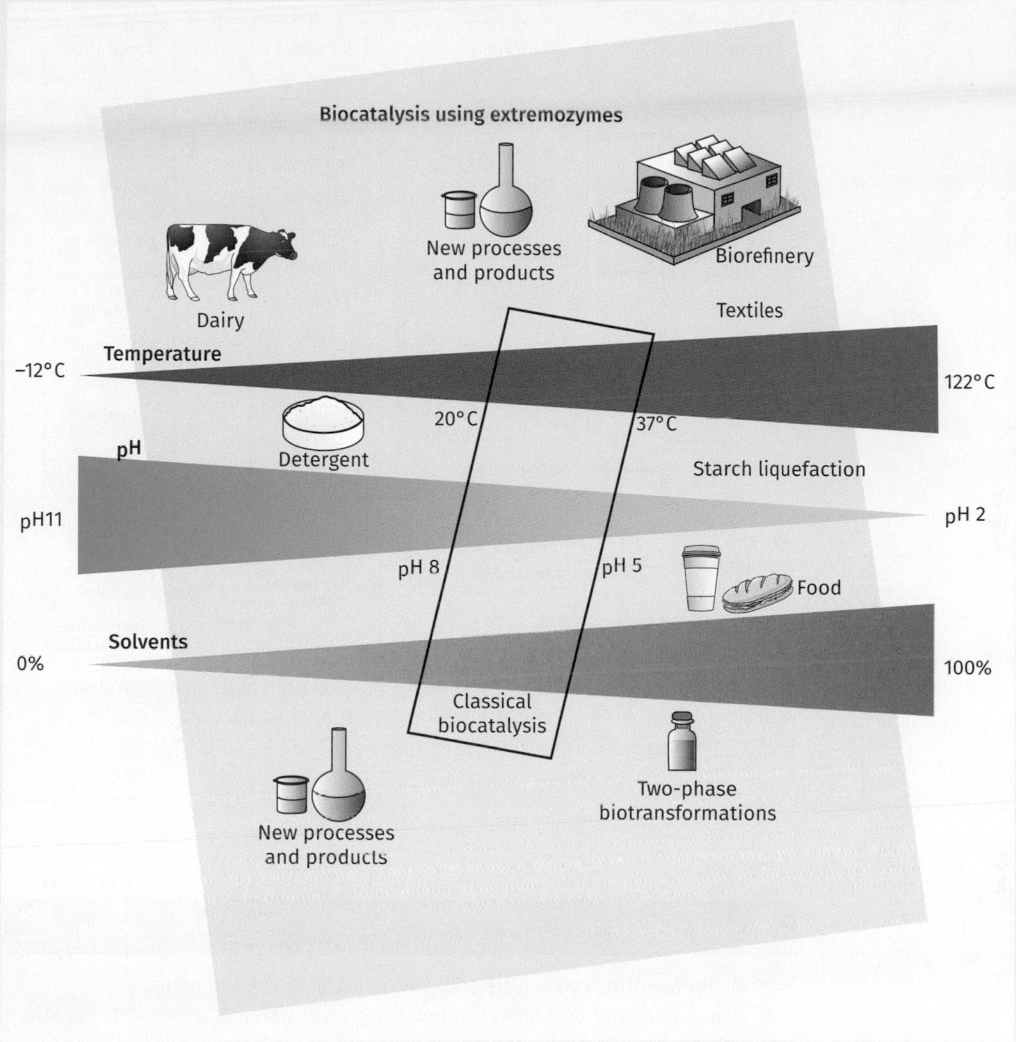

Adapted with permission from Elleuche, S., et al., Extremozymes—biocatalysts with unique properties from extremophilic microorganisms. Current Opinion in Biotechnology, 29, 116–123.

(e.g. as hosts), take tools out (e.g. enzymes; as depicted in Figure 9.2) or mimic processes (e.g. desiccation). Collecting samples from these habitats is challenging and resource-intensive, and even after a sample has been collected it may be impossible to support growth of the isolates in the laboratory.

Nucleic acids, proteins, and lipids of extremophile microorganisms need to be stable and functional at extreme growth conditions. Export systems, for instance, keep the intracellular conditions normal in chemical extremes, and the extreme environment requires extracellular enzymes to be stable. Proteins are temperature-stable due to highly charged surfaces, multiple ion-pairs, tight **hydrophobic core,** and generally more densely packing. Moreover, highly effective protein and DNA-repair systems allow microorganisms to live in extreme habitats.

› Fortunately, we have *in silico* approaches, e.g. culture-independent techniques and genome mining (remember Chapter 3).

Figure 9.2 Extremozymes can be used to process various sustainable feedstock materials. The diagram shows how this provides an input for the production of industrial chemicals.

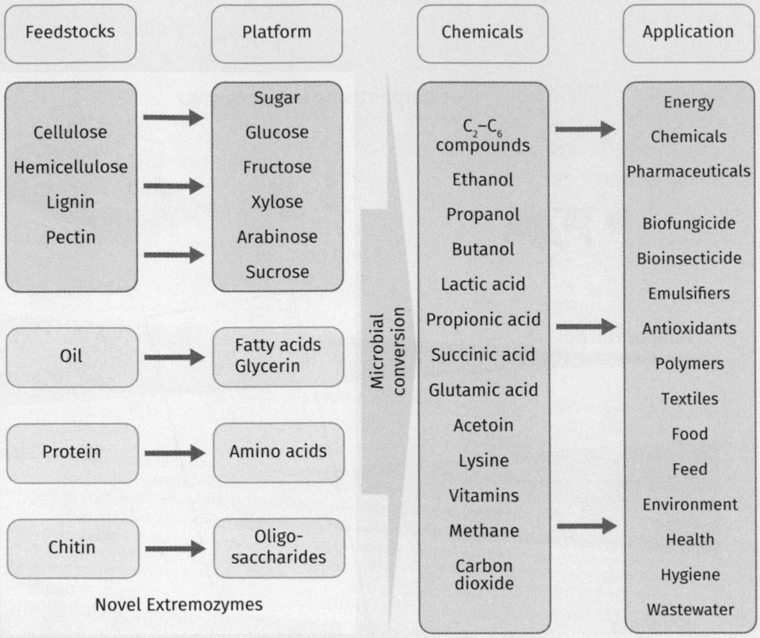

Adapted with permission from Elleuche, S., et al., Extremozymes—biocatalysts with unique properties from extremophilic microorganisms. Current Opinion in Biotechnology, 29, 116–123.

Key points

- Microorganisms can survive and grow in extreme environments.
- A minimum and maximum level of any growth parameter supports growth, and growth is fastest at the optimum level.
- Biological molecules, processes and structures are adapted to extreme conditions.
- Extremophile adaptations have applications in biotechnology.

9.1 Extreme temperature

At minimum growth temperatures, membranes almost gel and all transport processes and biochemical reactions are very slow. By contrast, at maximum temperatures, proteins are at risk of denaturing, membranes can collapse and general thermal lysis can occur. Optimal growth temperature supports enzymatic reactions at maximum rate. Adaptations throughout evolution enabled microorganisms to inhabit environments from below freezing point to above the boiling point of water, giving rise to diversity as shown in Figure 9.3.

Figure 9.3 Classification of organisms based on their optimal growth temperature. The terminology used for the organisms is given above the growth curves.

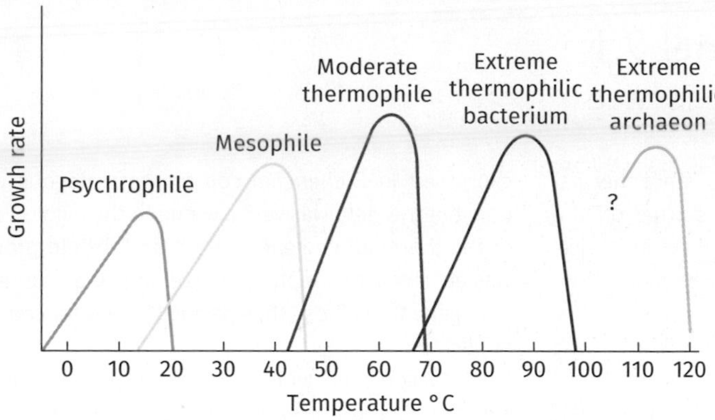

Heat

Extreme thermophiles can grow above 75°C, while moderate thermophiles grow at 55–75°C.

Thermoactive enzymes

Enzymes from extremophiles are also called extremozymes, and thermoactive enzymes (thermozymes) are functional in extreme conditions such as required for industrial processes. Enzymatic reaction rates roughly double whenever the temperature rises by 10°C. Thus, using thermozymes increases productivity and shortens process times.

› We have already covered some applications in bioproduction in Chapter 3 and will focus here on innovative examples in this context.

The success of PCR amplifications was made possible by various thermostable DNA-dependent DNA polymerases from a range of thermophilic organisms. Most of these enzymes have particular characteristics. The most widely used polymerase in PCR reactions is *Taq* (from *Thermus aquaticus*), but *Pfu* polymerase (from the hyperthermophilic archaeon *Pyrococcus furiosus*) is even more thermostable.

› These thermozymes have driven scientific advancements and still do, although in the meantime we have developed isothermal amplification without the need for a step at high temperatures (as discussed in Chapter 7).

Many thermostable DNA restriction endonucleases have also been isolated, for instance from *Thermus* species and *Bacillus stearothermophilus*. An extremely thermostable restrictase *Psp*GI was obtained from *Pyrococcus* sp. *Psp*GI can operate at 65–85°C and is an ideal component in DNA-based diagnostic applications.

Thermozymes and their novel applications are not only limited to DNA. The hydrolysis of plant cellulose for biofuel production is catalysed at high temperatures by lignocellulases from hyperthermophiles expressed in transgenic mesophilic plants once they have been harvested, while the thermozymes are inactive at growth temperatures of the plant.

The hyperthermophilic dehalogenase from *Sulfolobus tokodaii* has an optimal activity at 70°C but little at 25°C. Its activation can be triggered remotely by rapid heating of the enzyme to 70°C via immobilizing the enzyme on Fe_3O_4-based magnetic nanoparticles and applying an alternating magnetic field. This on/off switch for thermozymes has a range of applications for processes in challenging and inaccessible places such as the degradation of halogenated hydrocarbons and other pollutants.

Scientific approach panel 9.1
Thermus aquaticus, at your service

The natural habitat of *Thermus aquaticus* is hot thermal springs of 70–75°C. Using its polymerase started off the PCR market with a PCR kit and the Thermal Cycler device. Initially, the **Klenow fragment** of DNA polymerase I from *Escherichia coli* was used, which only operates at 37°C and denatures at temperatures required for PCR. Having to add fresh enzyme after every PCR cycle is time consuming, labour intensive, introduces contaminations, and limits the length of the generated PCR fragments to 400 bp. The *pol* gene from *Thermus aquaticus* was cloned and expressed in *Escherichia coli* for heterologous production, but the yield was very low due to the high GC content of the thermophile gene. More than ten-fold production was achieved by adapting the codon use of the recombinant gene to the host, thus paving the way for commercial applications.

Try to come up with as many as possible technologies that would not exist without this kind of PCR. (Re-) Familiarize yourself with alternative PCR approaches covered in this book.

Thermophilic α-carbonic anhydrase from *Thermovibrio ammonificans* can catalyse the hydration of carbon dioxide to bicarbonate. Overproduction in *E. coli* offers opportunities for commercial carbon dioxide capture.

Temperature-sensitive promoters

Promoters used for heterologous protein expression usually require an induction (e.g. via an inducer molecule), the addition or depletion of a nutrient or the change of a parameter such as pH. These inductions have disadvantages. Chemical inducers (e.g. IPTG, antibiotics) are expensive and need to be removed from the final product or process wastes to avoid toxicity.

> Thermoregulated expression systems are scalable (revise the challenges related to upscaling bioprocesses as discussed in Chapter 1), but limitations to heat transfer in large bioreactors need to be considered.

Alternatively, we can use heat-inducible expression systems based on thermophilic promoters, operated by increasing the temperature above 37°C.

Moreover, product and host need to cope with the change of temperature. The heat-shock response is controlled by the alternative sigma factor σ32 leading to quick synthesis of heat-shock proteins (stabilizing and repairing the folding of other proteins) and proteases (removing non-functional proteins). The growth rate also temporarily slows down. Again the success of process development lies in finding a suitable compromise of protein production and stress.

Such a compromise also involves considering the potential impact of temperature increase on recombinant plasmid segregation. For instance, some aspects of DNA replication in the frequently used heterologous host *Streptomyces* is temperature sensitive. Using two serial continuous cultures solves the problem. Biomass is produced in the first chemostat at low enough temperatures to repress the promoter and achieve a stable plasmid copy number. The outflow of this chemostat is pumped into the second chemostat, operated at the higher temperature needed for production of the recombinant protein.

Key points

- Thermoactive enzymes increase productivity and shorten process times.
- PCR amplifications were made possible by thermostable DNA-dependent DNA polymerases such as *Taq* (from *Thermus aquaticus*).
- Enzyme activity can be regulated via temperature changes when using thermozymes.
- Gene expression can be regulated (e.g. for heterologous proteins) via temperature changes.

Cold

Psychrophilic microorganisms, able to grow between −20 and 10°C but not above 15°C, dominate cold habitats, due to their ability to adjust membrane fluidity at low temperature, as well as to transcribe, translate, and catalyse biochemical reactions. Psychrotolerant organisms optimally grow at 20–25°C.

Cold-active enzymes

Most cold-active enzymes originate from psychrophiles and psychrotolerant microorganisms, but some have also been obtained from mesophiles and thermophiles. Their high activity below 25°C allows for sustainable and cost-effective industrial processes, as well as innovative applications in temperature-sensitive contexts.

Almost three-quarters of the world's population is lactose intolerant, creating a vast market for lactose-free dairy products. These can be produced by lactase (e.g. from *Kluyveromyces lactis*). Any temperature increase would affect the flavour of the milk and allow pathogens to grow. Psychrophile enzymes can be active at low temperature from production to storage at the consumer's home. Undesirable changes in taste and nutritional value can be avoided. A minimal rise in temperature inactivates these enzymes as required without any use of chemicals. The 'cleaning-in-place' of equipment using thermolabile enzymes therefore saves energy and avoids spoiling food due to residually active enzymes.

The universal blood type for transfusion therapy can be generated from type B blood by cold-active α-galactosidase from *Pseudoalteromonas* sp. It removes the antigenic component from surface carbohydrates of type B erythrocytes.

Cold-adapted enzymes are also invaluable in the cleaning and detergent industry. Lowering the temperature of the wash cycle from 40 to 30°C reduces electricity consumption per wash by 30 percent, i.e. 150–300 g of carbon dioxide emission. Over the course of one year this is equivalent to the emissions of about three million cars in Europe. Lowering the temperature further from 30 to 20°C, cuts carbon dioxide emissions by half, altogether lessening the impact on global warming.

Enzymes used in organic solvent mixtures, such as in the chemical synthesis industry, require a stable enough water hydration shell around their protein structure to support flexibility. Cold-adapted enzymes have a high **intrinsic flexibility** because of many hydrogen bonds between surface residues and solvent molecules, which at low temperatures offsets any effects of low viscosity. Low-temperature bioremediation of environments that are contaminated with various pollutants (e.g. during oil spills even in the Arctic), as well as the clearing of explosive and volatile compounds are further applications of cold-active enzymes.

Temperature-sensitive promoters

Using cold-inducible promoters allows gene expression at low temperature, which can enhance the solubility, functionality, and/or secretion of heterologous proteins. Some *E. coli* promoters are upregulated if the temperature is lowered from 37°C, but introducing promoters from psychrophilic organisms (e.g. *Pseudomonas syringae*) can improve productivity further, particularly in combination with the use of RNA polymerases that can transcribe at low temperature.

Cold-shock expression vectors containing the promoter from CspA (major cold shock protein in *E. coli*) add to the cold-active recombinant engineering toolbox. However, at low temperatures overall process productivity in *E.coli* is low. This can be remedied by using psychrophilic bacteria (e.g. *Pseudoalteromonas haloplanktis*) as expression hosts.

Freeze tolerance

Cellular functions and integrity are affected at low temperatures due to changes in water viscosity, membrane fluidity, diffusion rates, enzyme kinetics, and macromolecular interactions in general. Psychrophiles have evolved to successfully counteract the stress related to cold temperature (e.g. high osmotic pressure, desiccation, radiation, high or low pH, scarce nutrients). An overview of adaptations is provided in Figure 9.4.

Figure 9.4 Physiological adaptations of psychrophilic prokaryotes. A prokaryote cell is depicted with descriptions and locations of adaptations to cold temperatures.

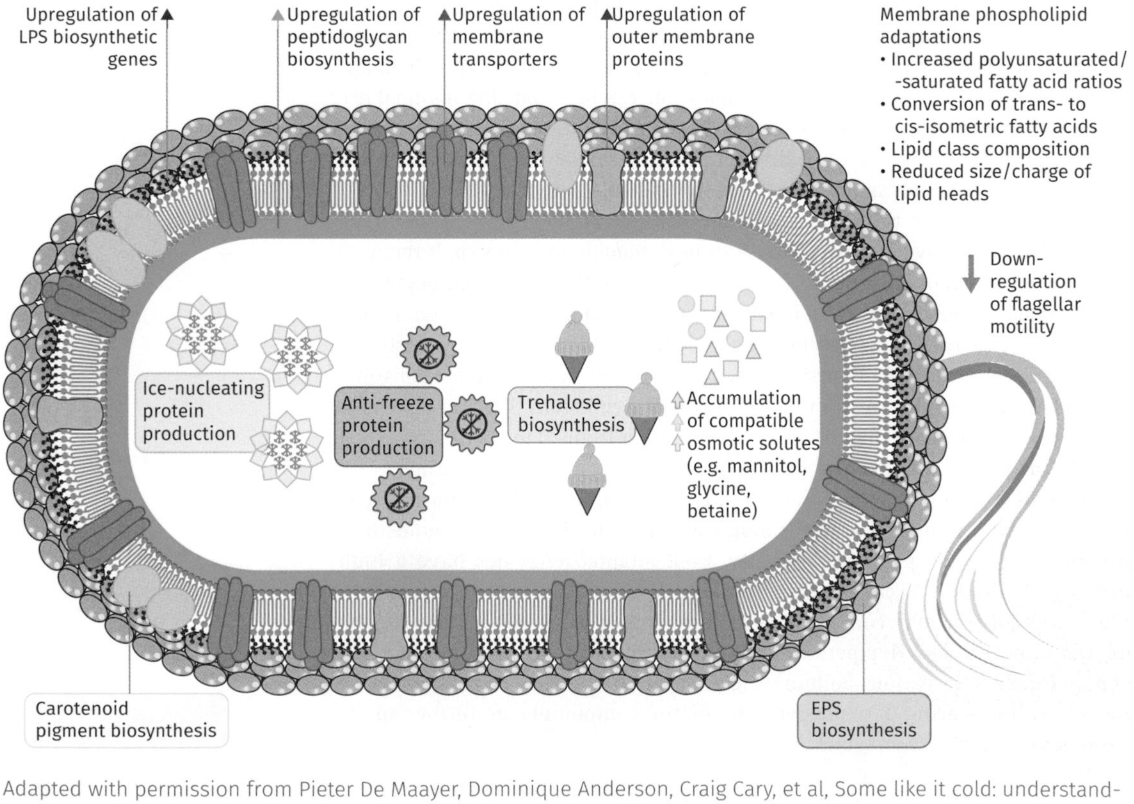

Adapted with permission from Pieter De Maayer, Dominique Anderson, Craig Cary, et al, Some like it cold: understanding the survival strategies of psychrophiles, EMBO Reports, John Wiley and Sons. Mar 26, 2014. © 2014 The Authors.

Dehydration is a considerable threat in frozen habitats. Bacteria accumulate small compounds (e.g. glucose, trehalose) to prevent themselves freezing, as well as producing ice crystal-controlling materials. This is possible because the freezing temperature of water depends on its homogeneity. If water contains additional particles such as dust or indeed ice-active bacteria, these act as seeds for ice nucleation and the solution can freeze at temperatures up to −2°C. Ice-nucleation at temperatures above −3°C is facilitated by ice nucleating proteins, whereas antifreeze proteins bind the surface of ice crystals and inhibit their growth.

If cytoplasmic ice crystals were to form, cellular damage depends on freezing rate and crystal location. Flash-freezing limits potential cryo-damage from intracellular ice. Extracellular ice can result in fractured membranes. The freezing temperature of external water can be lowered by secreted cryoprotectants. Intracellular cryoprotectants increase internal osmotic pressures prior to freezing. Moreover, the accumulation of compatible solutes (e.g. betaine, glycine, sucrose, mannitol, trehalose), triggered by an osmotic imbalance, lowers the cytoplasmic freezing point. Sugars stabilize both membranes and proteins too. This way the cells are protected against freezing and subsequent thawing, and more broadly against hyper-osmolarity and desiccation.

Antifreeze proteins, initially found in *Pseudomonas putida*, can be applied in the cryopreservation of biological materials (oocyte and erythrocyte storage) and cryosurgery tissue preservation. Frozen food quality and shelf-life can be increased using antifreeze protein treatments. For example, fermentation abilities of yeast in frozen dough is maintained, the smooth texture of ice cream is achieved, and drip loss in frozen meats is reduced.

Ice-nucleating bacteria are frequently Gram-negative plant pathogens (e.g. *Erwinia carotovora*, *Erwinia herbicola*). Such bacteria can be used as bio-insecticides. Recombinant *Enterobacter cloacae*, expressing an *Erwinia ananas ina* gene, can increase the freezing temperature of insect larvae following ingestion. Larvae then freeze upon exposure to temperatures of around −5°C. Ice nucleation proteins are also being used to produce artificial snow (as shown in Figure 9.5).

Figure 9.5 Production of artificial snow using ice nucleation proteins. A snow cannon produces snow for skiing.

Jason Finn/Shutterstock.com

Key points

- Cold-active enzymes allow for sustainable and cost-effective industrial processes and applications in temperature-sensitive contexts (therapeutic biological products, food industry, environmental bioremediation).
- Using cold-active enzymes in the cleaning and detergent industry cuts carbon dioxide emissions, lessening the impact on global warming.
- Antifreeze proteins can be applied in the cryopreservation of biological materials.
- Ice-nucleating bacteria can be used as bioinsecticides.

9.2 Extreme pressure

Oceans cover roughly 70 percent of the Earth's biosphere. At an average depth of 3,800 m, the pressure amounts to 38 MPa and temperatures are within the range of 1–3°C. Deep-sea microorganisms have evolved to thrive in these conditions. If they grow optimally above atmospheric pressure, we refer to them as piezophilic. The term previously used is barophilic. Figure 9.6 shows the range of pressure-adapted organisms.

There are habitats, such as **hydrothermal vents** in the sea floor, that combine high pressure and temperature, requiring marine microorganisms to withstand temperatures from 1–300°C and pressures from 0.1–110 Mpa, and to grow at some intersections. Lipids are the most pressure-sensitive biological molecules. The membranes of piezophilic organisms are more fluid, partly due to a higher ratio of unsaturated to saturated lipids, and pressure changes affect cell integrity and transport processes. Pressure also affects protein structure, stability, and function. Multimeric proteins such as ribosome subunits may dissociate at increasing pressure. Pressures from 300 MPa even cause monomeric proteins

Figure 9.6 Classification of organisms based on their optimal growth pressure. The terminology used for the organisms is given next to the growth curves.

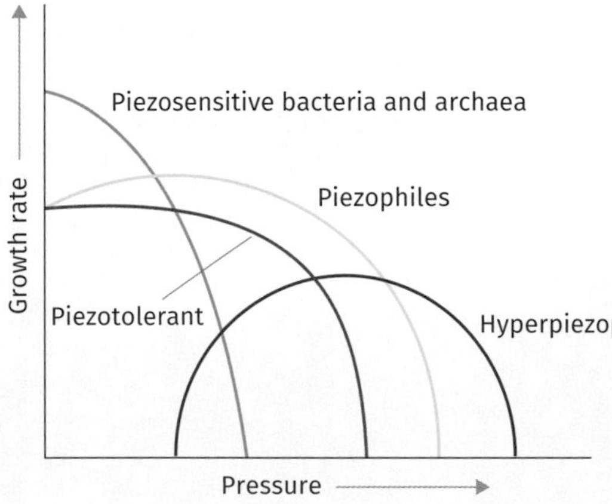

Reprinted from Trends in Microbiology, Sep 1, 2010, Jiasong Fang,Li Zhang, Dennis A. Bazylinski, Deep-sea piezosphere and piezophiles: geomicrobiology and biogeochemistry, Pages No., Copyright 2010, with permission from Elsevier.

to denature. Conversely, altering the pressure is a promising tool when counteracting protein misfolding. Pressure-induced sterical changes alter enzyme substrate specificity. This means we can control the product of an enzymatically catalysed reaction by applying certain pressure. The change depends on the magnitude of the activation volume and whether it is negative or positive. For example, a negative volume change is related to the hydrolysis of an anilide by chymotrypsin. Therefore, increasing the pressure promotes this reaction. However, the hydrolysis of an ester is related to a positive volume change, and increasing the pressure would inhibit the reaction. This concept can be widely applied in biotechnology, e.g. developing piezophilic strains of *Actinomyces* could lead to the production of novel antibiotics. Gene expression can also be regulated by pressure. The activity of the *E. coli lac* promoter, commonly used for recombinant protein production, can be considerably enhanced at 50 MPa even in the absence of an inducer such as IPTG.

Key points

- Altering the pressure can counteract protein misfolding or stabilize functional protein.
- Pressure-induced sterical changes alter enzyme substrate specificity.
- Gene expression can be regulated via pressure changes.

9.3 Extreme pH

As already mentioned, microorganisms can survive at pH values from 0 to 12, and grow within that range along the varied optimum values as shown in Figure 9.7. The pH value (i.e. hydrogen ion concentration) affects charge and dissociation of biological molecules and therefore their availability for any biochemical reactions. Undissociated organic compounds are also often toxic. The adaptations of microorganisms to function in extreme pH conditions allows their use in harsh industrial applications.

Figure 9.7 Classification of organisms based on their optimal growth pH. The terminology used for the organisms is given above the growth curves.

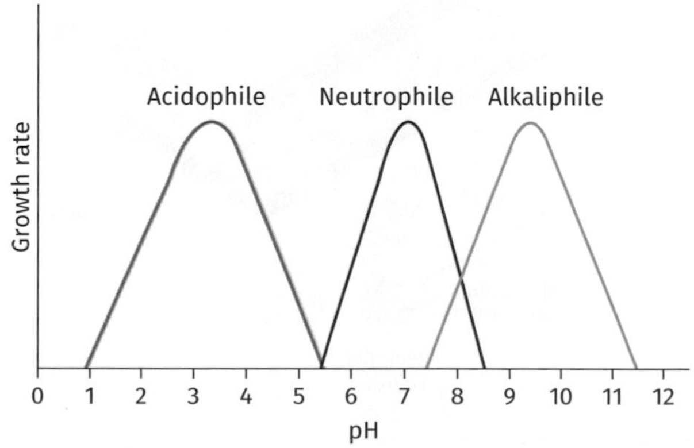

Biomining (microbiological leaching)

Microbial leaching dates back to the early 1950s and uses microorganisms to chemically convert solid metal compounds in waste materials to a soluble form. Such organisms are sulphur-oxidizing and iron-oxidizing bacteria that produce energy aerobically by chemolithotrophy, for example the acidophiles *Acidithiobacillus ferrooxidans*, *Acidithiobacillus thiooxidans*, and *Leptospirillum ferrooxidans*. Thermophilic species are also used such as *Sulfolobus* and *Metallosphaera* and even fungi such as *Penicillium* and *Aspergillus niger*.

The industrial process (see Figure 9.8) can be run in a heap, to which nutrients are added. The necessary air is provided by perforated pipes which also allows heat removal. The solution containing soluble forms of metal is continuously harvested for purification and metal recovery.

This biotechnical process of extracting metals is much safer and more environmentally friendly than traditional chemical heap leaching using toxic chemicals, and requires less energy and landfill space. Biomining achieves extraction rates of around 90 percent, whereas chemical leaching only results in 60 percent. Not only iron is being extracted as previously described, but many more metals—even uranium, and not only from ores. *Acidithiobacillus ferrooxidans* extracts lithium and copper from lithium batteries. Precious metals (silver, gold, and platinum) are extracted from catalytic converters of cars, jewellery waste, and electronic scrap using *Chromobacterium violaceum*, *Pseudomonas fluorescens*, and *Pseudomonas plecoglossicida*. The large volumes of electronic scrap (such as printed electronic circuit boards as seen in Figure 9.9) arising from the modern technologies in our lives contain valuable metals that must not be wasted. The components of a mobile phone contain about 28 percent metals, which include 10–20 percent copper, 1–5 percent lead, 1–3 percent nickel, and a

Figure 9.8 Diagram of a biomining process. The main components and flow between the processes is depicted.

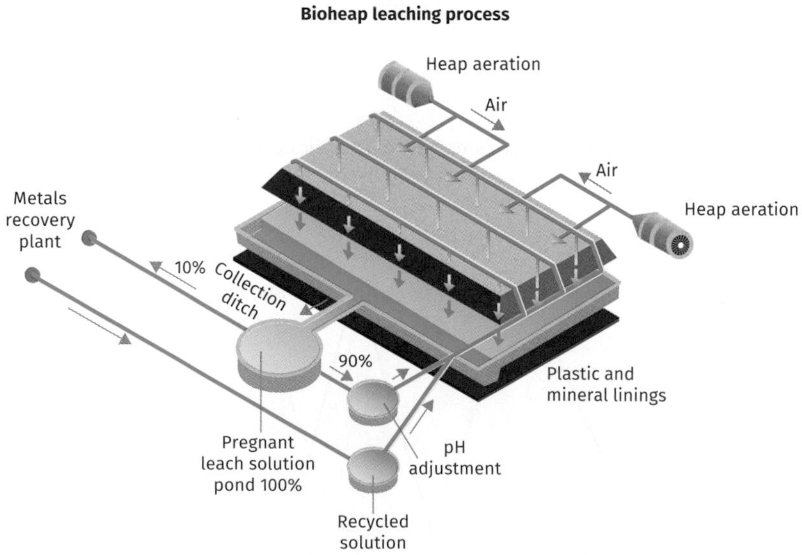

Figure 9.9 Electronic scrap. This picture shows that 'scrap' is a considerable resource thanks to microorganisms.

Jirik V/Shutterstock.com

further 0.3–0.4 percent comprising silver, platinum, and gold. *Acidithiobacillus ferrooxidans* and *Acidithiobacillus thiooxidans* can extract these metals, as well as aluminium and zinc.

Acidithiobacillus ferrooxidans produces $Fe_2(SO_4)_3$, which then oxidizes, e.g., solid copper in the scrap, resulting in the soluble copper ion.

The chemical reactions are as follows:

$$Cu + Fe_2(SO_4)_3 \rightarrow Cu^{2+} + 2Fe^{2+} + 3SO_4^{2-}$$

$$2FeSO_4 + H_2SO_4 + 0.5O_2 + \text{bacteria} \rightarrow Fe_2(SO_4)_3 + H_2O$$

Biomachining

The principle of 'solubilizing' metal is also applied in biomachining, where bacteria (e.g. *Acidithiobacillus thiooxidans, Acidithiobacillus ferrooxidans*) are used as renewable cutting tools. High-quality copper and mild steel parts can be cut with a tolerance of just 5 micrometres over 10 m length, where such accuracy is essential. Figure 9.10 shows the reactions involved. Bacterial culture or supernatant can be used. The finishing is smooth at low bacterial concentrations and becomes coarser at higher cell densities. Even engraving/etching of complex structures is possible. As illustrated in Figure 9.11, surface areas that should not be cut are covered with a protective layer, which is removed once cutting is finished.

Desulphurization of coal

A different technical application arises if the reactions focus on sulphur rather than on the metal involved. The sulphur compound pyrite (FeS_2) is a frequent inorganic component in coal. That is why burning coal pollutes the air with carbon dioxide and sulphur dioxide, leading to acid rain contaminating the

Figure 9.10 Diagram of biomachining a copper part. The bacterial cell and locations of relevant reactions are provided.

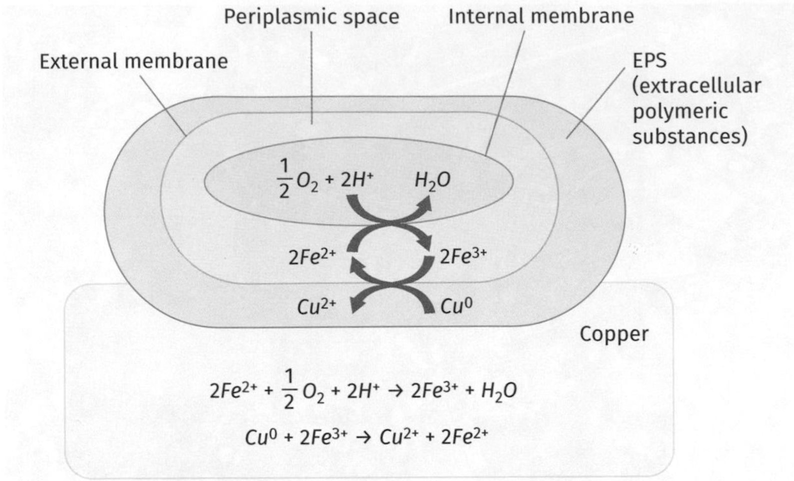

Figure 9.11 Diagram of biomachining and engraving. The main steps are illustrated.

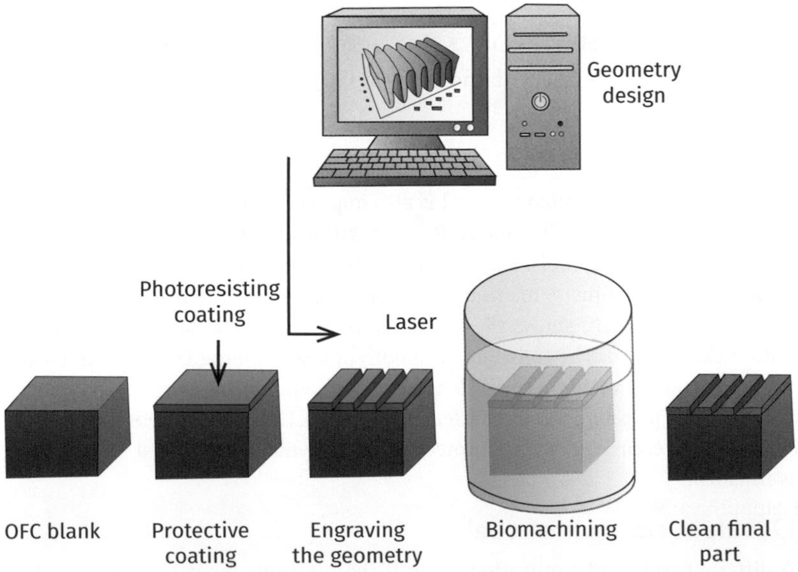

environment. This effect can be limited if sulphur is removed from coal before burning by solubilizing the pyrite.

Thiobacillus ferrooxidans oxidizes the sulphur at optimal temperatures of 28–30°C and optimal pH 1.8–2.2. If the bacteria are directly in contact with the material, the following reactions are involved:

$$FeS_2 + 3,5\,O_2 + H_2O \rightarrow FeSO_4 + H_2SO_4$$
$$2FeSO_4 + 0,5\,4O_2 + H_2SO_4 \rightarrow Fe_2(SO_4)_3 + H_2O$$

Pyrite is oxidized to iron(II) sulphate and sulphuric acid, which further react with oxygen to iron(III) sulphate and water.

Alternatively, the bacteria are indirectly employed by producing a substance that then oxidizes the minerals. At low pH the active substance is Fe^{3+}, which contributes to the following reactions:

$$Fe_2(SO_4)_3 + FeS_2 \rightarrow 3\,FeSO_4 + 2S^0$$
$$2S^0 + 3O_2 + 2\,H_2O \rightarrow 2\,H_2SO_4$$

The iron(III) sulphate is soluble and its Fe^{3+} is reduced to Fe^{2+} while the sulphide is oxidized to elemental sulphur, which further reacts to become sulphuric acid.

Enzymes

Extracellular enzymes from acidophiles have to function at low pH. Many extremozymes (e.g. amylases, proteases, cellulases) isolated from acidophiles are active under harsh industrial conditions such as when processing starch or fruit juice, or producing animal feed and pharmaceuticals. Stability at low pH and any other extreme conditions cannot be narrowed down to a certain factor. It is due to a combination of factors (e.g. intraprotein interactions, enhanced ion pairing or hydrophobic interactions, reduction in the surface area to volume ratio), which have been optimized during evolution. Alkaliphile enzymes, isolated from microorganisms growing optimally at or above pH 9, are being commercialized e.g. in the leather industry, where hide-dehairing is conducted at pH 8 to 10.

› Alkaliphile enzymes are frequently used in detergents as discussed in Chapter 3.

Key points

- Microbial leaching uses sulphur-oxidizing and iron-oxidizing microorganisms to chemically convert solid metal compounds in waste materials to a soluble form.
- The principle of 'solubilizing' metal is also applied in biomachining.
- A similar process is used for removing sulphur from coal to limit contaminating the environment.
- Enzymes from acidophiles and alkaliphiles are widely commercialized.

9.4 Extreme osmotic pressure

Halophilic microorganisms thrive in the presence of salt, and extreme halophiles can survive up to 30 percent (w/v) salt (about 5 M NaCl) concentration. Figure 9.12 shows the range of osmo-adapted organisms.

One adaptation is the 'salt in' strategy, whereby the salt concentration inside the cell is increased to balance osmotic pressure. Significant changes to protein features allow functionality at such high salt concentrations. Usually, halophilic

Figure 9.12 Classification of organisms based on their optimal growth osmotic pressure. The terminology used for the organisms is given in the legend.

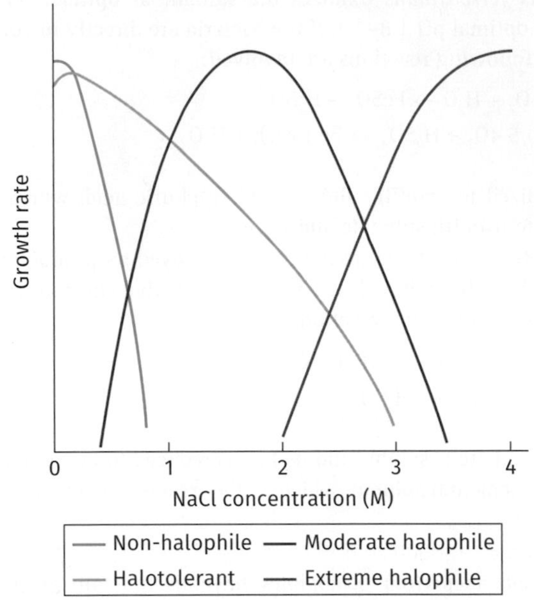

proteins have an excess of acidic amino acids on their surface to prevent aggregation. Alternatively, osmotic pressure is balanced by accumulating inorganic ions (usually KCl), thus preventing NaCl diffusion into the cell.

Most halophilic organisms accumulate low molecular weight compounds referred to as compatible solutes (i.e. osmolytes, extremolytes). These soluble organic compounds can make up to a quarter of dry cell weight when exposed to stressful conditions.

Compatible solutes

Compatible solutes protect biological molecules and thus cells from damage. For instance ectoine, hydroxyectoine, and glycine betaine generally stabilize biological structures. This means the cells are adapted to salts and desiccation, heat and cold (including freezing conditions as mentioned in section 9.1). Ectoines were the first extremolytes to be manufactured on a large scale for applications such as cell protectants in skin care, or for stabilizers of proteins and cells. Such stabilization also includes protection of recombinant proteins against misfolding, degradation, and aggregate formation. Ectoine (Figure 9.13) and hydroxyectoine are zwitterions and thus have a strong ability to bind water.

Figure 9.13 Chemical structure of ectoine.

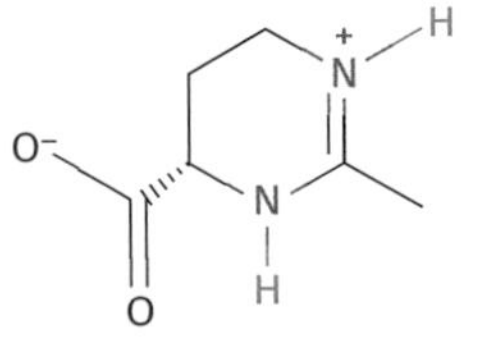

The stabilizing effect is frequently explained by the 'preferential exclusion model', which means that the compatible solutes are excluded from the hydration shell of a biological macromolecule (e.g. protein), as shown in Figure 9.14. This results in a preferential hydration of the surface of that macromolecule, which enhances the stability of the functional conformation, and denaturation becomes less favourable thermodynamically.

Figure 9.14 Illustration of the preferential exclusion model. The solute is surrounded by either cosolvents or water.

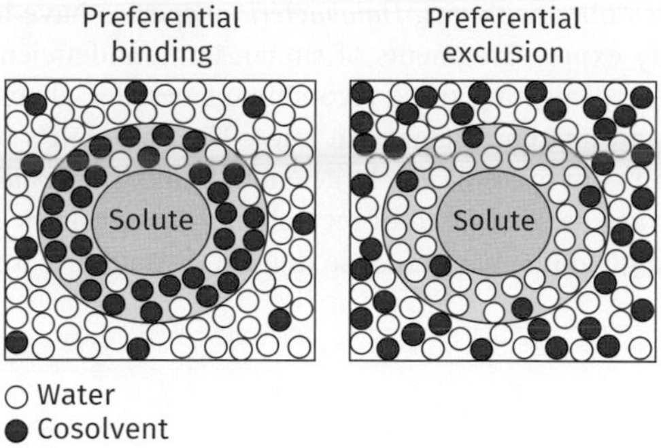

Adapted from S. Moelbert, B. Normand and P. de los Rios, Kosmotropes and chaotropes: modelling preferential exclusion, binding and aggregate stability, Biophys. Chem. 112, 45 (2004).

The large-scale production of ectoines uses a technology called 'bacterial milking' based on applying an osmotic downshock to the culture. *Halomonas elongata* is grown to a high cell density in a medium containing 2.57 M NaCl. Cells are then filtered and the medium is replaced with distilled water. The cells rapidly release ectoine to recover the intracellular osmolarity. The ectoine-containing water is removed, and salty medium is added again to restore hyperosmotic conditions.

Non-halophilic microorganisms such as *E. coli* can be engineered to produce ectoine, but engineering halophiles themselves allows for higher productivity.

Carbohydrates such as trehalose (e.g. produced in *Sulfolobus solfataricus*) are also compatible solutes and can be applied commercially, for example to stabilize antibodies and vaccines.

Pharmaceutical and cosmetic products frequently draw on osmolytes as ingredients.

Several UV-protective compounds (ectoine, scytonemin, mycosporin-like amino acids, bacterioruberin) originate from extremophiles and are used in sunscreens. The radioresistant microorganisms *Halobacterium* and *Rubrobacterium* produce bacterioruberin, which has been suggested for the prevention of UV-induced skin cancer such as melanoma. The compound contributes to the repair of UV-damaged DNA.

Products from halophiles

Halophiles produce various products of commercial interest. Amylases from, e.g., *Halomonas*, *Halobacillus*, and *Streptomyces* are used in the food industry. Proteases for laundry detergents and other applications are manufactured by *Bacillus*, *Halobacillus*, and *Chromohalobacter*. *Streptomonospora* and *Halomonas* produce xylanases and cellulases for biobleaching. Halophiles are also a source for biosurfactants and bioemulsifiers which are amphiphilic molecules that allow homogeneous mixing of hydrophobic and hydrophilic substances. Even plastics for medical materials can be manufactured using halophiles (e.g. using *Haloferax mediterranei*).

Alternative vaccine delivery

Some halophilic microorganisms produce internal gas vesicles which are small protein structures filled with gas. *Halobacterium* species have been engineered to recombinantly express fragments of simian immunodeficiency virus on the gas vesicle surface. Injecting these recombinant vesicles elicits a strong antibody response and immune memory in mice. The halophile's own polar lipids can act as an adjuvant to result in a large enough immune response. Even a nasally delivered vaccine, composed of recombinant gas vesicles and polar lipids, has resulted in a good immune response in mice without any toxicity.

Case study 9.1

The two faces of extremophiles

In 1956 the extremophile *Deinococcus radiodurans* (Figure CS 9.1) was isolated from X ray-irradiated tinned meat, having survived due to unparalleled efficiency when repairing DNA damage. *Deinococcus* species are known for resistance to multiple stresses (oxidation, desiccation, radiation, 4–55°C), posing a challenge as contaminants in harsh industrial processes. *Deinococcus geothermalis* frequently causes problems in paper machines because it forms **biofilms**. It anchors itself to abiotic surfaces and neighbouring cells via protein adhesion threads. These biofilms require **pulsed polarization** to detach them from stainless steel surfaces. But we can also use *Deinococcus* species constructively in biotechnological applications such as bioremediation. Bioprocesses can also be improved by transferring the unique resistances to recombinant hosts. Additionally, bioproducts from deinococci have various applications. Deinoxanthin, a carotenoid isolated from *Deinococcus radiodurans*, can induce apoptosis in cancer cells and therefore be potentially used as a chemopreventive substance.

Consider what such extremophile adaptations mean for our approaches to disinfection/sterilization. If microbes can resist radiation, think about potential survival in space.

Figure CS 9.1 Electron microscopic image of *Deinococcus radiodurans*. The image illustrates the cell morphology.

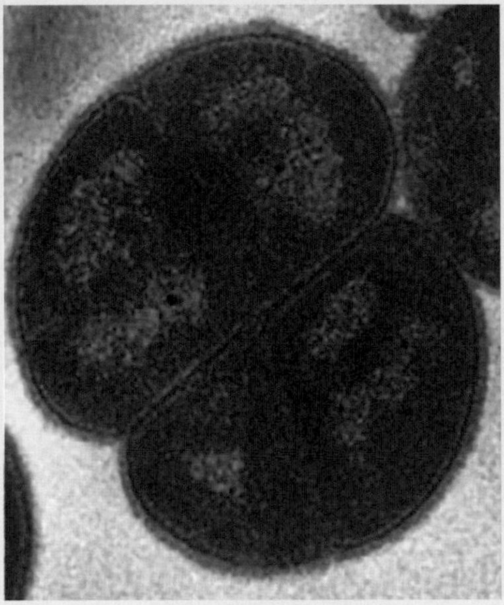

Michael J. Daly, Uniformed Services University of the Health Sciences, Bethesda, MD, USA

The bigger picture panel 9.1

The conservation of microbial ecosystems

When collecting samples from remote, extreme environments that have thus far been isolated from human reach, we pose a risk to their integrity by introducing contamination in the form of pollutants and 'foreign' mesophilic microorganisms. Such contamination may alter the ecosystem and its microbial diversity. We could lose new microbial know-how in the actual attempt of finding and isolating it. One example of such a process is non-aseptic deep ice drilling as illustrated in Figure BP 9.1. What are the issues in this complex situation, and are there any that have not yet been alluded to? What are the challenges in addressing these issues?

Figure BP 9.1 A diagram of a deep ice drilling operation. Visualizing how we access pristine habitats gives food for thought.

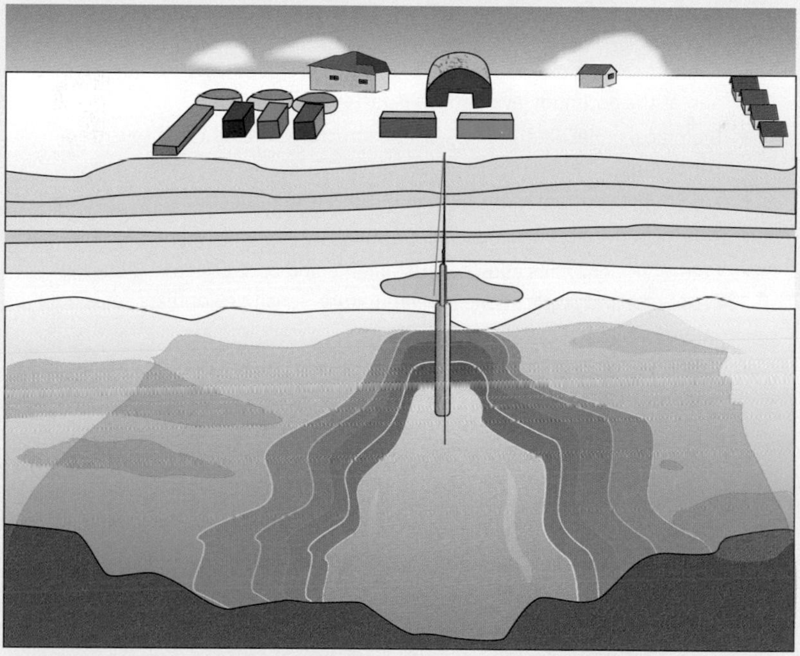

Science History Images/Alamy Stock Photo

Key points

- Osmolytes protect biological molecules (including recombinant proteins) and entire cells from damage. This is commercially applied.
- Osmolytes can be produced large-scale via 'bacterial milking'.
- UV-protective compounds from extremophiles are used in sunscreens and could prevent UV-induced skin cancer.
- Halophiles produce useful enzymes, biosurfactants, and bioemulsifiers.
- Internal gas vesicles are an option for alternative vaccine delivery.
- Extremophiles can be responsible for contaminating and impairing industrial processes.

Chapter Summary

- Microorganisms can survive and grow in extreme environments.
- A minimum and maximum level of any growth parameter supports growth, and at the optimum level growth is fastest.
- Biological molecules, processes, and structures are adapted to extreme conditions.
- Extremophile adaptations have applications in biotechnology.
- Thermoactive enzymes increase productivity and shorten process times.
- Cold-active enzymes allow for sustainable and cost-effective industrial processes and applications in temperature-sensitive contexts, lessening the impact on global warming.
- Antifreeze proteins can be applied in the cryopreservation of biological materials.
- Altering the pressure can counteract protein misfolding or stabilize functional protein.
- Microbial leaching uses sulphur-oxidizing and iron-oxidizing microorganisms to chemically convert solid metal compounds in waste materials to a soluble form.
- Enzymes from acidophiles and alkaliphiles are widely commercialized.
- Osmolytes protect biological molecules (including recombinant proteins) and entire cells from damage. This is commercially applied.
- UV-protective compounds from extremophiles are used in sunscreens.
- Halophiles produce useful enzymes, biosurfactants, and bioemulsifiers.
- Internal gas vesicles are an option for alternative vaccine delivery.
- Extremophiles can be responsible for contaminating and impairing industrial processes.

Further Reading

Ishino, S. and Ishino, Y. (2014). 'DNA polymerases as useful reagents for biotechnology—the history of developmental research in the field'. Front. Microbiol 5: 465.
This paper provides a historical perspective of DNA polymerase developments.

Krüger, A. et al. (2018). 'Towards a sustainable biobased industry—Highlighting the impact of extremophiles'. New Biotechnology 40: 144–53.
This paper reviews extremophiles in the context of sustainability.

Raddadi, N. et al. (2015). 'Biotechnological applications of extremophiles, extremozymes and extremolytes'. Appl. Microbiol. Biotechnol. 9: 7907.
This paper reviews extremozymes and extremolytes as applications.

Yin, J. et al. (2015). 'Halophiles, coming stars for industrial biotechnology'. Biotechnology Advances 33: 1433–42.
This paper reviews the context of halophiles.

Discussion Questions

9.1 Discuss the difference between survival and growth in extreme environments.

9.2 Debate why your favourite extremophile application in biotechnology is more beneficial than any other.

9.3 Think about how climate change could affect microbial habitats.

10 MICROBIAL BIOTECHNOLOGY IN THE ART AND BUILT ENVIRONMENT

Learning Objectives

- To be able to explain the impact of microorganisms on our cultural, urban, and industrial heritage;
- to be able to describe the relevance of microbial food webs and succession to biodeterioration;
- to be able to relate technologies that minimize microbial growth, to prevention of biodeterioration;
- to be able to provide an overview of novel sustainable materials and tools based on microorganisms.

We may be used to seeing some of our cultural, urban, and industrial heritage in ruins due to activities of nature (e.g. earthquakes) or human action (e.g. wars), but what may come as a surprise is the constant destructive impact of microorganisms. From buildings to sculptures (see Figure 10.1), paintings and pipelines, our art and **built environment** is a microbial habitat providing an abundance of substrates. For instance, stones are minerals, oxides, and salts (e.g. limestone is calcium carbonate). Microorganisms do occur naturally in these habitats or are introduced by humans. Environmental pollution adds to the problem. **Chemolithotrophic** bacteria (e.g. **nitrifiers**) corrode limestone by producing readily soluble nitrate, which replaces less soluble carbonate in the stone. **Sulphur- and iron-oxidizing** bacteria oxidize sulphur compounds or Fe^{2+}, damaging concrete pipelines via sulphuric acid. **Chemoheterotrophs** also dissolve mineral components and leach or chelate cations due to organic acids or secreted polymers. Even marble is slowly but steadily dissolved by citric acid from *Aspergillus versicolor*; as is basalt by *Candida albicans*. The idiom 'constant dripping wears away a stone' comes to mind. (For a deeper understanding of microbial nutrition read the Microbial Physiology primer in this series.) Further damage in the form of discolouration or pigmentation is caused because of microbial growth (e.g. green patinae due to cyanobacteria; black spots due to fungi). Other organisms from lichens on stone to invertebrates (e.g. termites on wood) are also responsible for damage and destruction.

Understanding the complex interactions and activities of microbial communities that cause biodeterioration allows us to develop and select optimal preservation approaches and technologies. This follows the same principles as taking a sample from a patient—isolating and analysing the causative agent to decide on the treatment. Where microorganisms cannot be grown, next-generation sequencing techniques are used to determine the composition of the microbiome, and to identify the mechanisms of the destructive activities. Prevention and remediation methods can then be chosen using evidence to stop or at least limit microbial growth, to clean and rescue our cultural heritage and civic and industrial assets. Interestingly, these methods can actually involve using microorganisms. Moreover, microorganisms themselves can be used in the creative industry and in art.

10.1 Biodeterioration

The phenomenon of biodeterioration is described in the Bible (Leviticus 14: 36), there referred to as red and green leprosies. This shows how long we have been observing structural and aesthetical damage of our environment due to acid, mechanical, and enzymatic destruction.

Every year hundreds of billions of USD have to be spent worldwide to protect artwork in museums alone. However, if such books and paintings were to be irreparably damaged, the cultural loss would far extend any monetary value. Damage to industrial buildings and housing adds further to the economic burden. The microorganisms involved can also be pathogens; and indoor microbial pollution is directly linked to poor public health. It is therefore crucial to develop and use microbiologically resistant construction materials. The so called 'green building materials' such as bamboo can actually worsen the situation because of their ability to retain water, which supports fungal growth.

Stone

Microorganisms cause weathering and corrosion (as shown in Figure 10.2.) when growing on stone surfaces and in its pores which contain water. Acid rain as well as other pollutants, particularly in urban environments, increase the biodeterioration by changing the chemical parameters of the ecosystem and/or by adding carbon sources and other nutrients to it.

Figure 10.1 Degradation of stone heritage. Microbial growth, staining, and stone damage is visible.

Peter Turner Photography/Shutterstock.com

Built environment

We find a succession of organisms and a food web in the built environment like in any other ecosystem. When colonizing stone, photolithoautotrophic microorganisms support other microorganisms.

Nitrifiers (e.g. *Nitrosomonas*, *Nitrobacter*) produce stone-corroding nitrous and nitric acid from ammonia. Sulphur oxidizers (e.g. *Thiobacillus*) cause damage by producing sulphuric acid from reduced sulphur, and chemoorganotrophs (e.g. *Bacillus*) contribute with organic acids. The natural carbonates of stone are unstable in acid solutions, and metal (e.g. iron, magnesium, calcium) cations are leached from stone minerals by these acids. Chemical changes then lead to physical changes. Stone becomes more brittle and can then be invaded by fungi and lichens, worsening the damage. Microfractures and increased porosity facilitate pollutant build-up and further biodeterioration. Microbial

Figure 10.2 Context of biological colonization of stone as an ecosystem. The environmental growth factors and interactions within the microbial food web are shown.

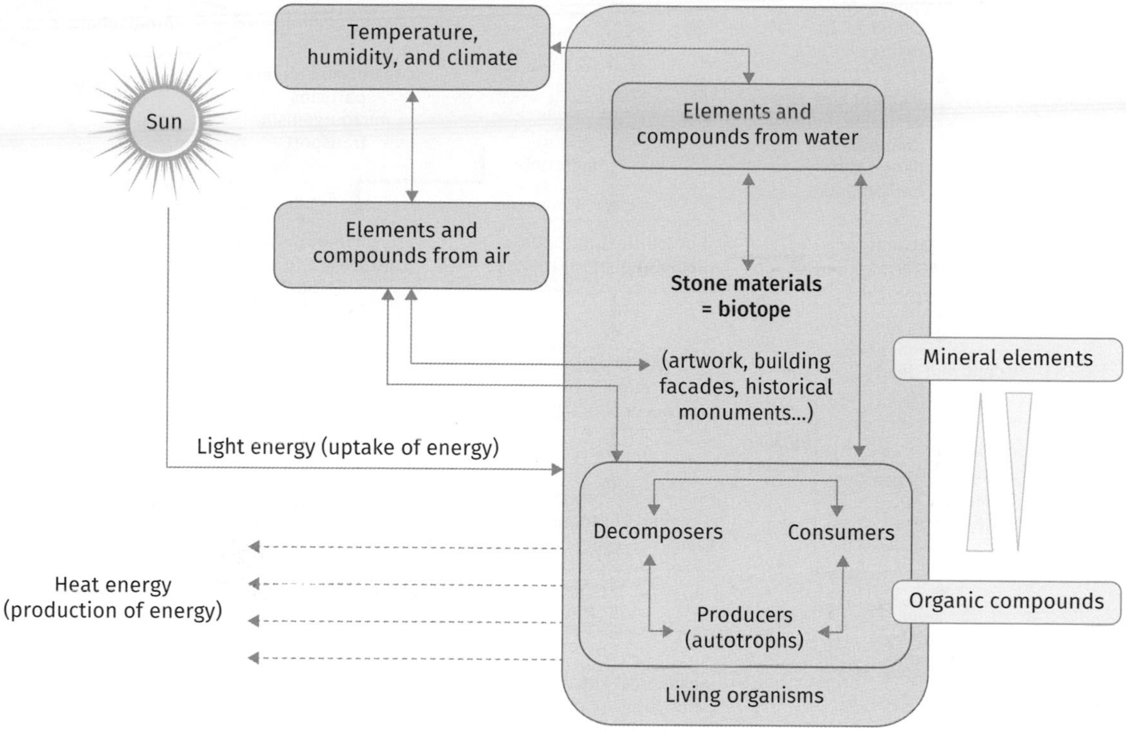

Adapted with permission from Fabien Borderie, Badr Alaoui-Sossé, Lotfi Aleya, Heritage materials and bio-fouling mitigation through UV-C irradiation in show caves: state-of-the-art practices and future challenges, Environmental Science and Pollution Research. Springer Nature, Jan 1, 2014. Copyright © 2014, Springer-Verlag Berlin Heidelberg

growth and activity (including biofilm formation) increases surface humidity and related damage. The by-products of microbial metabolism or cell pigmentation cause discolouration. The complexity of biodeterioration is summarized in Figure 10.3.

It is worth looking at concrete biodeterioration in more detail. Concrete's wide use means its damage impacts on the economy. Mixed biofilm communities of anaerobic sulphate reducers and facultative anaerobes are the most corrosive. *Desulfovibrio* sp. are the most common sulphate reducers in oil field constructions which use a lot of concrete. Biogenic sulphuric and other acid corrosion is probably the most common cause for underground pipelines to fail, nearly as frequently as in sewage systems.

Fresh hydrated cement has a high pH (around 13) due to calcium hydroxide formation, which limits microbial activity. Over time, concrete reacts with carbon dioxide and is neutralized by volatile hydrogen sulphide (abundant in sewers, wastewater treatment plants, and buildings that house animals) lowering the surface pH to around 9. These conditions allow microbial colonization and succession (see Figure 10.4) with *Thiobacillus thioparus* at the start of the process. *Thiobacillus thioparus* oxidizes thiosulphate, resulting in a nearly

Figure 10.3 Overview of biodeterioration pathways on stone. Biological activities and impact on stone are outlined.

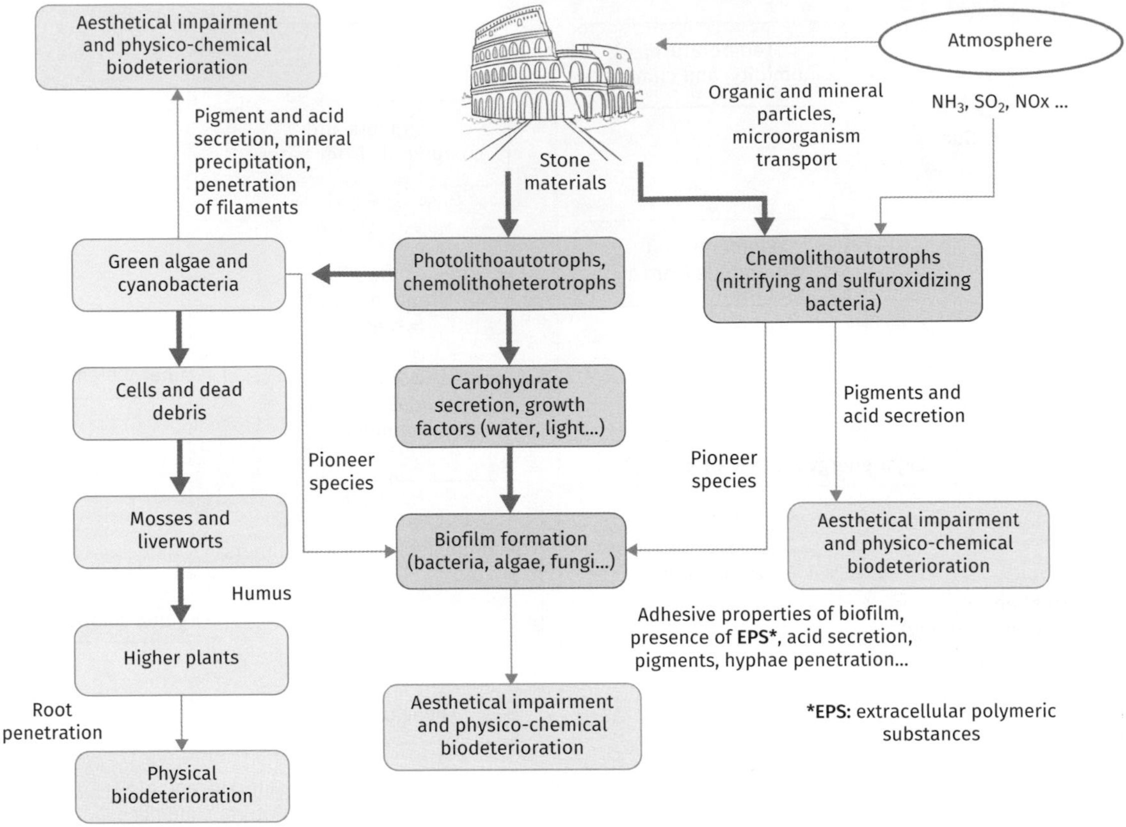

Dakal, T.C., Cameotra, S.S. Microbially induced deterioration of architectural heritages: routes and mechanisms involved. Environ Sci Eur 24, 36 (2012) doi:10.1186/2190-4715-24-36

neutralized surface pH, allowing additional species to grow and thereby further lowering the pH. Below pH 5, thiosulphate is converted to sulphur, fuelling the sulphur oxidation pathway and *Acidithiobacillus thiooxidans* can flourishe. The pH will eventually be as low as 1.5 due to additional sulphuric acid, limiting microbial growth but severely corroding concrete. Biogenic sulphuric acid can react with calcium hydroxide to form gypsum (calcium sulphate dihydrate). A gypsum layer can protect concrete, but if it gets washed away the exposed surface quickly corrodes. Microorganisms very readily penetrate even seemingly intact concrete. This enhances surface wear, particularly in frost periods, and can alter concrete features (e.g. heat transfer).

Art

All the materials discussed in this chapter and more are found in art and are subject to respective biodeterioration processes.

In this section we will look at Lascaux Cave (France) as a prominent example of biodeterioration. The sealed cave was discovered in 1940 and as it had not been accessed for 17,000 years the prehistoric art of the Upper Palaeolithic period was well preserved (see Figure 10.5).

Figure 10.4 ***Thiobacillus* sp, succession on fresh concrete exposed to hydrogen sulphide.** The cumulative activity of the different species increases corrosion over time.

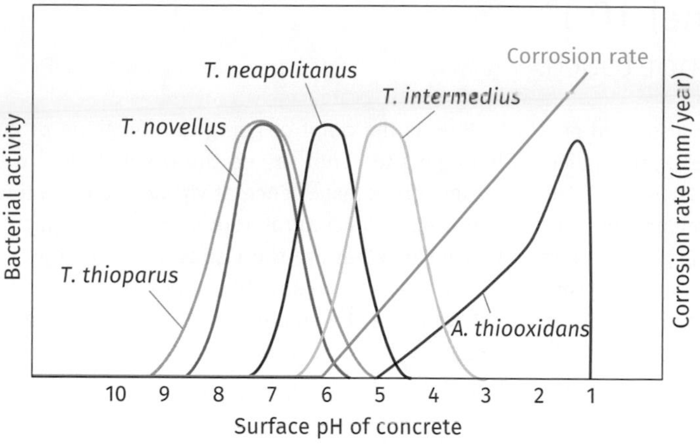

Adapted from Islander et al. Journal of Environmental Engineering. 117: 751–770. With permission from ASCE. Any other use requires prior permission of the American Society of Civil Engineers. This material may be found at https://doi.org/10.1061/(ASCE)0733-9372(1991)117:6(751)

Figure 10.5 Lascaux Cave paintings.

Hemis/Alamy Stock Photo

The pigments used in the paintings as well as the stone 'canvas' are an excellent microbial substrate. Unsurprisingly, the find attracted scientists and endless streams of hundreds of visitors per day. These visitors brought moisture, carbon dioxide, heat, pollutants, and microorganisms. Lights were installed for better visibility, which supported phototrophs (e.g. algae). Due to significant microbial deterioration, the cave was closed in 1963 to limit pollution and allow for preservation. A replica of the cave was opened for visitors.

A succession of outbreaks referred to as 'sicknesses' from green to white to black took place over time due to fungi (e.g. *Fusarium solani*, *Chrysosporium*,

The bigger picture panel 10.1

Immerse yourself, but do not touch.

Art and culture should be accessible, but also need protecting. How to square that circle? Lascaux II, the underground replica was opened in 1983, and in 2009 Lascaux III, a replica of the paintings followed as a travelling exhibition. Lascaux IV, a full replica of cave and paintings, opened in 2016. These efforts have been made in the interest of conservation, and copies of famous paintings are frequently displayed in museums for insurance purposes.

What is the point of going to a museum or cultural heritage site if not to see the original? And with the all-immersive experience of virtual reality, will we see a decrease in cultural tourism? What impact would that have? What are the issues in this complex situation, and are there any that have not yet been alluded to? What are the challenges in addressing any of the issues?

Gliocladium, Trichoderma, Verticillium), algae (e.g. *Bracteacoccus*), and bacteria (e.g. *Pseudomonas fluorescens*). Biocide treatments, including those with quaternary ammonium compounds and even antibiotics, have been conducted at various times and as a result selected for pathogens (e.g. *Legionella*, *Escherichia*). Some isolates are also biocide resistant (e.g. *Ralstonia*, *Pseudomonas*). The example of the Lascaux Cave strongly makes the case for informed treatment to target the cause, as will be discussed later.

Key points

- Cultural, urban, and industrial heritage is being destroyed by microorganisms.
- Understanding microbial interactions and activities that cause biodeterioration allows optimal preservation approaches and technologies.
- Succession of microorganisms and a microbial food web occur like in any other ecosystem.
- Photolithoautotrophs support other microorganisms when colonizing stone and forming biofilms.
- Nitrifiers produce stone-corroding nitrous and nitric acid; sulphur oxidizers damage by sulphuric acid, and chemoorganotrophs by organic acids.
- Natural carbonates of stone are instable in acid solutions, and metal cations are leached from stone minerals by these acids.
- Microbial metabolic by-products or cell pigmentation cause discolouration of the material.
- Treatments can cause further damage or select for resistant organisms.

Metal

Roughly 20 percent of all corrosion damage is attributed to microorganisms directly or indirectly. Like with concrete, biodeterioration of steels and **alloys** causes disruption to industries and economic loss. In buildings, once the concrete layer has been destroyed, steel reinforcements can be attacked.

Figure 10.6 Microbiological influenced corrosion of metal. The serious damage to the fragment is visible as a change of colour.

Again, sulphate Remreducers are active. Corrosions shown in Figure 10.6 can be anaerobic (e.g. *Desulfovibrio*) or aerobic (*Thiobacillus*), with the latter causing swift damage by creating conditions of up to 10 percent sulphuric acid. The acid reacts with the metal to form salts, which are soluble, and the metal corrodes over time.

Additionally, **electrochemical corrosion** takes place. The corroding metal surface serves as a mixture of **cathodes** and **anodes**, and any water present acts as **electrolyte**. The electrochemical reactions and destruction of the metal depends on the electrolyte pH, temperature, humidity, and oxygen availability.

Wood

We find wood in both art and the built environment. It is frequently destroyed by fungal growth, particularly in poorly ventilated environments with high humidity. The outcome is called rot, because the wood is rotting away.

There are three types of rot (as shown in Figure 10.7):

- Brown rot due to enzymatic activity and oxidation, usually results in high-strength losses in the wood. *Serpula lacrymans* is a cause, and infected wood needs to be removed. Potassium hydroxide can be used for prevention.
- White rot (e.g. caused by *Trametes versicolor*) results in less strength, which can be stopped by drying out the wood.
- Soft rot is due to cellulolytic enzymes (e.g. produced by *Trichoderma* sp.), that decompose hardwood. Chemical treatment can stop the decay.

Carriers of information

If buildings weather we can repair or rebuild them. But what if the building plan is irretrievably lost? We live in the information age and are used to readily having access to information at any time and anywhere. This was not always the case, and we are at risk of losing our information/data because of microorganisms.

Figure 10.7 **Wood decay due to (a) brown rot; (b) white rot; (c) soft rot.**

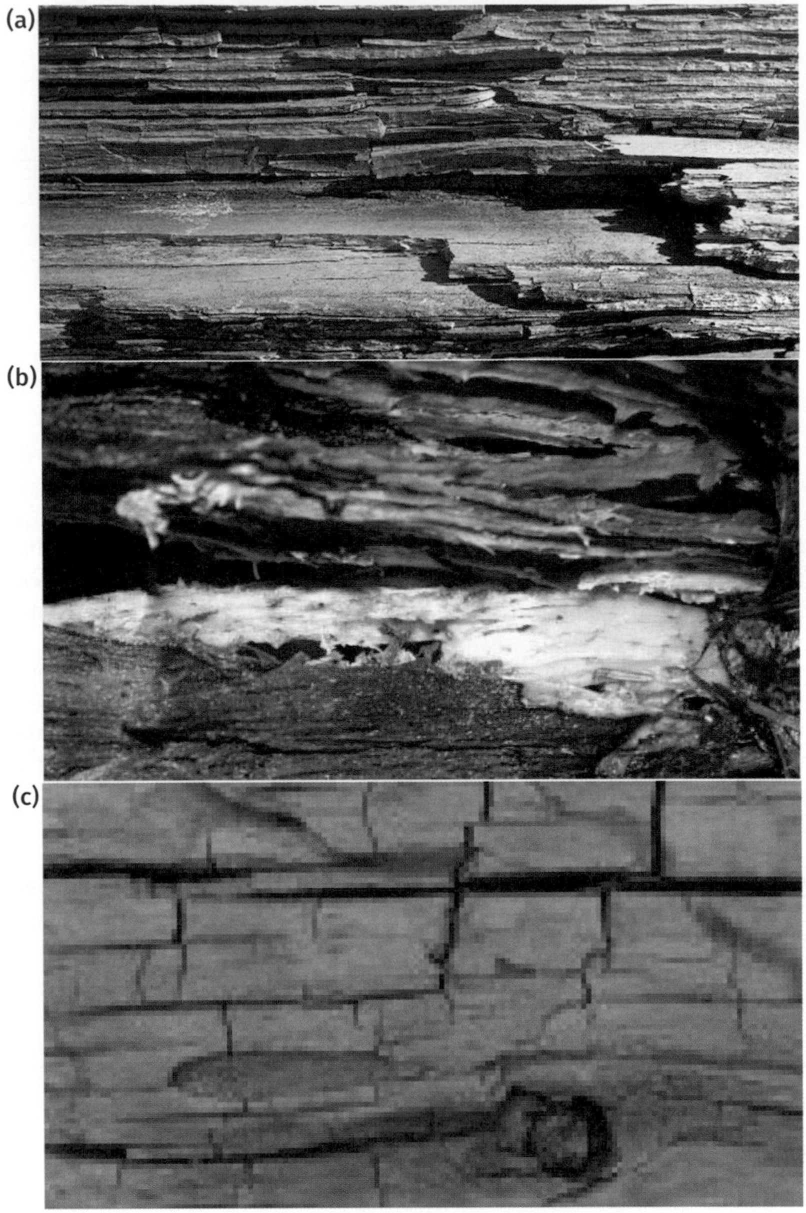

(a) and (b) Custom Life Science Images/Alamy Stock Photo (c) Yon Marsh/Alamy Stock Photo

Paper

Paper is a comparatively rich microbial habitat with high water activity containing cellulose, lignin, pectin, minerals, and proteins—supporting the growth of fungi and bacteria. Microbial enzymes (e.g. cellulases) damage paper and all the information on it. There is not just destruction to deal with but also the deterioration caused by staining or discolouration. The 'foxing' of paper due

Figure 10.8 Foxing spots on paper.

ChartsTable789/Shutterstock.com

to fungi results in brown spots as depicted in Figure 10.8, and can be used to authenticate paper-based artefacts.

In addition to biodeterioration, contamination of books with pathogens (e.g. *Staphylococcus* spp., *Aspergillus* spp.) is a concern. Library books can readily serve as vectors in the transmission of infectious diseases, commonly respiratory and skin infections. Historically, steam, carbolic acid, or formalin vapours have been used to disinfect the stock, which in themselves also pose health risks—more informed measures of disinfection are required.

Parchment

Collagen, keratin, and elastin are components of parchment. Biodeterioration is caused by oxidation of amino acid chains and hydrolysis of the peptides. Microorganisms also cause damage due to their acid discolouration of the parchment or their pigments staining it. The protein-based habitat supports growth of bacterial genera such as *Pseudomonas*, *Bacillus*, and *Staphylococcus*, and also fungal genera such as *Aspergillus* and *Trichoderma*.

Compact discs

Compact discs were introduced in the early 1980s to store binary information grooved as lands and pits into a polycarbonate layer. The discs contain aluminium. Hyphae of a *Geotrichum*-like fungus have been found in discs, as depicted

Figure 10.9 Deteriorated compact disc due to fungal growth. Hyphae are clearly visible.

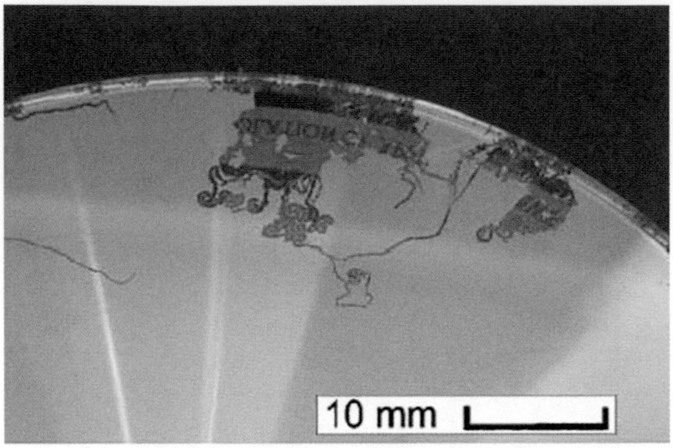

Reproduced with permission from Garcia-Guinea, J., Cárdenes, V., Martínez, A.T. et al. Naturwissenschaften (2001) 88: 351. https://doi.org/10.1007/s001140100249

› Although compact discs have been on the way out, our modern data storage facilities are subject to biodeterioration (as discussed for electronic scrap in Chapter 9). Microorganisms do not wait inactive until we declare something to be obsolete technology.

in Figure 10.9, which can solubilize the metal, grow through the layers and degrade them—destroying the records of digital data altogether.

Even 'the cloud' has its physical manifestation in the form of servers. If DNA computing becomes mainstream, our information will be at risk because microorganisms have the tools to degrade DNA.

Key points

- Biodeterioration of steels and alloys causes disruption to industries and economic loss.
- Metal corrosion by sulphate-reducers can be anaerobic or aerobic.
- Wood is frequently destroyed by fungal growth, causing brown, white, and soft rot.
- Paper is a comparatively rich microbial habitat, and microbial enzymes and pigments damage it.
- Biodeterioration of parchment is caused by oxidation of amino acid chains and hydrolysis of peptides.
- Compact discs contain polycarbonate, aluminium, and other components which serve as microbial nutrients.

10.2 Countering biodeterioration

› As an example, museums use single-cell biosensors (as we discussed in Chapter 7) for environmental monitoring and control of conditions in order to limit deterioration.

The more we know about the roles and mechanisms of microbial biodeterioration, the better we can prevent, preserve, and restore effectively without causing detrimental side effects to the object and the environment. Any such attempt starts with an anamnesis, like with a human patient, covering treatment history, evaluation of the value, and analysis of the damage and contributing factors, including identification of the microorganisms. We can then make a decision on how to address the damage, monitor the success of the treatment, and limit damage for the future.

Countering colonization

As soon as a microorganism attaches to a surface, biodeterioration commences, and the damage increases with the formation and build-up of biofilms. Naturally, killing and removing the microorganisms limits the damage, ideally without causing extra damage to the object but preventing regrowth. This problem is not new and the Phoenicians (1500–300 BC) combined mechanical removal of microbial growth with antibacterial compounds such as copper salts to preserve objects.

Nowadays, biocides are no longer used on a trial-and-error basis but on careful selection, particularly in the context of priceless artefacts. Frequently used in restoration are quaternary ammonium compounds, ethanol, formaldehyde releasers, and isothiazolinone. Detergents, oxidizing and chlorinated compounds, and aldehydes can be applied for less sensitive objects, but many are toxic and **ecotoxic**. Several countries have banned, e.g., ethylene oxide even though it is the most effective mass treatment against mould in libraries. Some nitrogen-based biocides can act as nutrients for subsequent colonization.

Physical methods provide an alternative to chemical ones, but again are not generally appropriate. UV light, for instance, does not penetrate enough and can affect pigments. Gamma radiation is effective against spores and fungi in doses of more than 0–20 kGy, but can alter materials. Laser techniques can be suitable. Biological cleaning agents are also available, and these can have a microbial origin. For instance *Desulfovibrio desulphuricans* and *D. vulgaris* (anaerobic sulphate reducers) are able to remove black sulphate crusts on stone.

Potential damage to the object is not the only limitation to countering microbial growth. The microbial communities are mixed, and different organisms in the community will respond differently to the approaches, making full removal difficult. Targeting a specific mechanism is therefore more promising than targeting specific organisms. Controlling **quorum sensing** can oppose biofilm formation on objects.

Biomineralization

Microorganisms do not only solubilize and destroy stone, but they also contribute to stone formation, and can be considered bioengineers, and employed to repair damage.

A range of bacteria promote carbonate precipitation (e.g. calcium carbonate) in the context of core metabolic processes that increase the pH in their microhabitat. This results in the concentration of carbonate, then supersaturation, and then precipitation. Bacterial cells are usually negatively charged and can also aid carbonate precipitation directly by attracting positively charged calcium ions (see Figure 10.10). Using the precipitation of calcium carbonate (e.g. marble and limestone) is an eco-friendly technology when restoring stone in the art and built environment.

The actual process depends on biotic (type of bacteria) and abiotic factors (salinity, nutrient availability, calcium concentration, pH). Bacterial bioengineers are found in four main categories:

- photosynthetic organisms;
- sulphate reducers;
- organisms reducing amino acids/nitrate or hydrolysing urea;
- organisms utilizing organic acids.

Figure 10.10 Bacteria as nucleation site for calcium carbonate precipitation. The negatively charged bacterial surface attracts calcium ions. Carbonate precipitation takes place after addition of urea due to release of inorganic carbon and ammonium as one example of biomineralization.

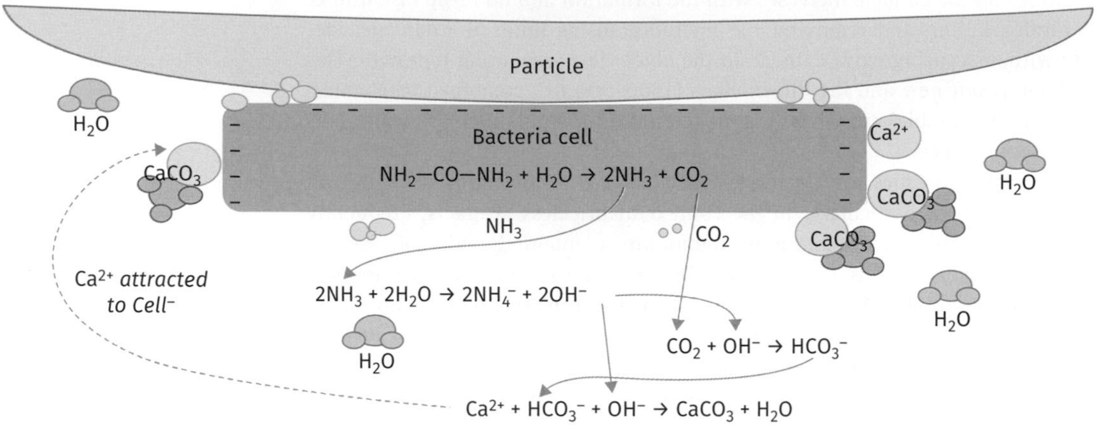

Net urea hydrolysis reaction: $NH_2—CO—NH_2 + 3H_2O \rightarrow 2NH_4^+ + HCO_3^- + OH^-$

Net pH increase: $[OH^-]$ generated from NH_4^+ production >> $[Ca^{2+}]$

Adapted from DeJong, J.T., et al., Bio-mediated soil improvement, Ecological Engineering, Elsevier, February 2010.

Microorganisms can also produce a protective surface on stones via oxalic acid, precipitating calcium as calcium oxalate.

Even geotechnical applications are possible—where soil is stabilized to protect it from erosion, or to facilitate tunnelling or to improve its bearing capacity. Initially, the precipitates decrease soil pore size and then bind soil particles together, gradually leading to denser and denser soil.

Scientific approach panel 10.1
Self-healing stone

Bacterial biomineralization of stone is a cost-effective approach to extending the lifetime of materials such as concrete—it works by sealing up microcracks. It is important to have viable bacteria present at the site of damage that become active when required. Including bacteria in the mixture is one option—growth will only commence once cracks, which automatically give access to nutrients and water, have developed. Precipitation of calcium carbonate then seals and heals the crack in an environmentally friendly and durable way (see Figure SA 10.1). Spore-forming bacteria such as *Bacillus* are robust enough to persist in the alkaline environment above pH 9 and guarantee long-term survival in the material. Alternatively, encapsulation can be used where bacteria are contained, e.g. in polymer microcapsules, which are mixed into the concrete. When cracks form and break such capsules the bacteria become activated.

Do you think this is a temporary solution, or could it protect stone for good? Do you think the occupants of the buildings might be concerned about having bacteria enclosed in their walls?

Figure SA 10.1 Figure Cracks in concrete in the process of healing using Bacillus sp.

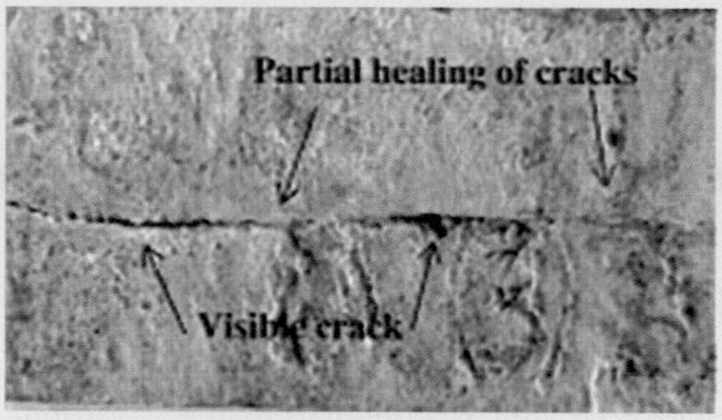

From Krishnapriya, S., et al., Isolation and identification of bacteria to improve the strength of concrete. Microbiological Research, May 2015, Copyright © 2015 Elsevier GmbH. All rights reserved

Prevention of deterioration

Where appropriate, biodeterioration should be prevented by minimizing growth supporting factors. That would include adjusting humidity and temperature (e.g. via climatization controls), as well as light in closed systems (e.g. housing, industrial buildings) and places where cultural heritage is found or kept. This is impossible outdoors so alternatives are required.

Nanotechnology can help, e.g. with silicon dioxide nanoparticles protecting surfaces against pollutants and microorganisms. Nanomaterials with novel features that self-protect and prevent corrosion have been proposed and developed with far more to come. These can be embedded in concrete or incorporated in surface treatments to control deterioration, or at least prolong durability. Large-scale manufacturing of the nanomaterials at low cost has yet to be achieved for routine application, and long-term safety tests are required. Surface treatment technology based on inorganic nanoclays looks very promising.

Key points

- Countering biodeterioration is optimized by specific informed actions.
- Killing and removing microorganisms limits damage.
- Careful use of biocides limits toxicity, development of resistance, and introduction of nutrients.
- Physical methods such as radiation provide an alternative.
- Microorganisms are used as biological cleaning agents.
- Biomineralization, the bacterial precipitation of calcium carbonate, is an eco-friendly restoration technology.
- Biodeterioration can be prevented by minimizing growth supporting factors, and by protecting materials using e.g. nanotechnology.

10.3 Creating and creativity

Applying our understanding of microbial activities allows us to prevent damage to and assist in the restoration of artwork including historical fabrics (e.g. affected by *Actinomyces dermatonosum*, *Pseudomonas fluorescens*, *Alternaria*, *Cladosporium*). Such understanding also gives us access to novel materials and tools for creativity while being environmentally friendly and sustainable, from production via consumption to disposal.

Materials for clothing

We are used to producing our clothes from plant-based (e.g. cotton, linen) and animal-derived (e.g. wool, silk, leather) material. But microorganisms have their place in fashion too.

❯ We can produce, e.g. cellulose fibres just from sweetened tea (as shown in Figure 10.11), with the help of yeast and bacteria, specifically *Gluconacetobacter xylinus*, which we already know from Chapter 4 in the context of producing the beverage kombucha.

The cellulose fibres fuse together to form a sheet (see Figure 10.12) to be used as pieces of material, and even these pieces can be fused together by mere overlapping and then drying without any sewing necessary. If the pieces are placed onto a garment form, the entire seamless item is moulded in one production step.

Variation to the production procedure, starting from tea, results in leather-like material suitable for anyone not wishing to wear animal products.

Figure 10.11 Production of cellulose fibres in a kombucha culture.

Oscar Burriel/Science Photo Library

Figure 10.12 Cellulose fibre sheet produced in a Kombucha culture, which will be washed and dried for further processing.

Lightenoughtotravel/Wikimedia Commons/Public Domain

Figure 10.13 Moisture-reactive vents in fabric. If the wearer sweats, vents are opened to cool.

The culture is kept at room temperature, just using kombucha, green tea, sugar, and vinegar. After a while a transparent skin forms on the surface, and within three to four weeks a thick layer of a couple of centimetres can be harvested. It needs to be carefully washed with cold water and some soap, then spread out and dried. It is then ready to be cut and processed like any other material.

In addition to making clothes from bacterial products, we can also incorporate bacteria into the material for functional clothing. An example is humidity-sensitive fabric as a result of *Bacillus subtilis natto* biofilm being printed onto fabric. The bacterium is harmless and is in fact edible. It is used in Japanese cooking in a dish called nattō. Every bacterium can increase its size by up to half in response to increased humidity. This ability is applied to operate cooling vents (Figure 10.13) in clothes. When the temperature of a body part increases, the wearer sweats and needs ventilation. Due to their change in size, the cells curl open flaps in the clothing which close again once the body has cooled down and stops sweating.

Material can also respond to heat and be manipulated to change colour. So lampshades can adjust to the light in the room and sunscreens on windows can change orientation and shape in response to the weather.

Dyeing fabrics

We have been using dyes from plants, animals, and minerals for thousands of years and artificial dyes more recently. Frequently, the dyes or the process itself cause health and environmental concerns. So again sustainable alternatives are sought and pigment-producing bacteria are available to replace traditional techniques.

Many of us enjoy wearing blue jeans and are familiar with the indigo dye, originally from the plant *Indigofera tinctoria*. We can produce indigo to dye jeans (Figure 10.14), using microorganisms in various ways. Genetically modified *Escherichia coli* express naphthalene dioxygenase from *Pseudomonas putida* to, in combination with its own tryptophanase, oxidize naphthalene. During production, isatin, an inhibitory compound of indigo synthesis, is reduced or removed from the culture broth to increase indigo production, as

Figure 10.14 Pairs of jeans dyed with microbially produced indigo.

well as improve the characteristics of the indigo. Alternatively, engineered *Escherichia coli* are able to synthesize indigo based on indole or tryptophan.

We can also use bacteria to dye fabrics directly or print patterns on fabrics (Figure 10.15). *Streptomyces*, for instance, are naturally rich in pigments and are already used to dye silk. The product can be influenced and driven by adjusting the culture conditions such as temperature, incubation time, volume, and pH. Potentially, silkworms could be genetically engineered to express specific microbial pigments to produce colourful silk.

Furniture from fungi

Microorganisms can even be used to produce furniture, and therefore offer natural and sustainable solutions for customers. The material is a mushroom composite which, unlike other composites, can be produced without toxins and carcinogens. This makes production, use, and disposal much safer. The material is toxin-free and actually edible. The process starts by inoculating a cheap carbon source such as agricultural wastes (e.g. corn husks) with mycelium of *Ganoderma lucidum*. Over a few weeks the fungus will grow, and the mycelium develops an extended fibrous structure. This fungal mass is then placed into a mould to take shape, while continuing to grow until the mould is fully packed. The fungus is then killed using heat, which denatures all proteins and stabilizes the composite. To finish the design, depending on the furniture, wooden legs are added, as can be seen in Figure 10.16. It is possible to adjust the stability and hardness of the final composite by varying the incubation parameters.

Figure 10.15 Dress 'Communicating Bacteria' dyed using bacterial pigments, by A. Dumitriu.

'Communicating Bacteria Dress' by Anna Dumitriu (photograph by Anna Dumitriu)

Figure 10.16 Stools produced using mushroom composite.

Lisa Ellsworth/Workshop Residence

Case study 10.1
True colours

Antimicrobial textiles are sought after for health and hygiene products, and metals (e.g. silver) have been used with fabrics for some time. Combining colours and antimicrobial protection is comparatively new. The phenazines pyorubin and oxychloraphin from *Pseudomonas* sp. have been characterized in their use. Silk dyed with these compounds are effective against a range of bacteria, e.g. *E. coli*. The more the fabric is dyed the better, and dye uptake has been optimized to 54 percent (pyorubin) and 48 percent (oxychlororaphin). Uptake is best at pH 3. Dying with pyorubin is completed after about an hour at 70°C for pyorubin, and after about 90 min at 90°C for oxychlororaphin. We can also achieve a range of shades for the same compounds by adding **mordants**. Without a mordant, pyorubin is bright red, but turns e.g. flame red with added copper sulphate, and terracotta with added iron(II) sulphate. Oxychloraphin is light yellow without mordant, but adding, e.g. iron(II) sulphate, turns it chrome yellow on silk.

Do you think such antimicrobial activity affects our normal flora? Consider whether other fabric dyes we are using might have antimicrobial activities.

Key points

- Microbial activities give access to novel, environmentally friendly and sustainable materials, as well as tools for creativity.
- Microorganisms can produce cellulose to be used in clothing.
- Incorporating bacteria into material allows producing functional fabrics.
- Pigment-producing (naturally or engineered) bacteria are available to replace traditional dyeing techniques.
- Fungi can be used as composite material in furniture.

Chapter Summary

- Cultural, urban, and industrial heritage is being destroyed by microorganisms.
- Understanding microbial interactions and activities that cause biodeterioration allows optimal preservation approaches and technologies.
- Succession of microorganisms and a microbial food web occur like in any other ecosystem.
- Microbial metabolic by-products or cell pigmentation can cause discolouration of material.
- Killing and removing microorganisms limits damage.

- Careful use of biocides limits toxicity, development of resistance, and introduction of nutrients.
- Physical methods such as radiation provide an alternative.
- Microorganisms are used as biological cleaning agents.
- Biomineralization, the bacterial precipitation of calcium carbonate, is an eco-friendly restoration technology.
- Biodeterioration can be prevented by minimizing any growth supporting factors and by protecting materials (e.g. using nanotechnology).
- Microbial activities give access to novel, environmentally friendly and sustainable materials, and tools for creativity.

Further Reading

Dhami, N. K., Reddy, M. S., and Mukherjee, A. (2014). 'Application of calcifying bacteria for remediation of stones and cultural heritages'. Frontiers in Microbiology 5: 304.

This paper reviews stone deterioration and applying bacteria for repair.

Loto, C. A. (2017). 'Microbiological corrosion: mechanism, control and impact—a review'. International Journal of Advanced Manufacturing Technology 92: 4241–52.

This paper reviews microbiological corrosion.

Rinaldi, A. (2006). 'Saving a fragile legacy'. EMBO Reports 7: 1075–9.

This paper discusses damage to cultural heritage and how biotechnology can be used to counteract it.

Sterflinger, K. and Piñar, G. (2013). 'Microbial deterioration of cultural heritage and works of art — tilting at windmills?' Applied Microbiology and Biotechnology 97: 9637–46.

This paper discusses the pros and cons of biocide treatments.

Discussion Questions

10.1 Discuss how environmental pollution can enhance biodeterioration.

10.2 Discuss the economic context of biodeterioration.

10.3 Discuss how biotechnology can contribute to sustainable materials.

11 ETHICAL CONSIDERATIONS

Learning Objectives

- To be able to give an overview of bioethics issues in the context of microbial biotechnology;
- to be able to explain how ethical considerations need to be approached;
- to be able to outline public perception of microorganisms and biotechnology;
- to be able to describe the context of biopatents;
- to be able to discuss ethical issues around bioweapons and bioterrorism.

Biotechnology allows us to develop tools and processes which benefit our and future generations—in all aspects of our life, from medicine to environment to food. Given this positive potential why do we have to think about ethics or regulation in biotechnology? And what can go wrong? We have already come across some broad and complex examples of bioethics issues in the bigger picture panels in other chapters. Bioethics concerns go beyond microbial biotechnology processes such as tissue engineering, transplantations, and **3D bioprinting**. There are frequent worries relating to genetically modified organisms, including microorganisms and crops. Genetic modification has many benefits, but threats to health, environment, confidentiality, access to research output, informed consent, and intellectual property are issues. Moreover, how different belief frameworks view changing the genetic information of an organism, or even constructing it fully synthetically, needs to be considered. Also, large-scale biotechnological production facilities (e.g. growing large volumes of engineered microorganisms) can be considered a risk and the relevant safety issues have to be addressed. We also have to think about aspects of labelling food or other products in terms of their origins.

In the case of patenting, issues relate to the technology as such, to innovation and ownership. While these concerns seem more remote, more obvious ones arise from the dangers of bioterrorism potentially using genetically engineered pathogens, or from how information and/or data (e.g. diagnostic information) can be accessed and used, and thus impact on individuals and society.

❯ We have already discussed some of these ideas in Chapters 3 and 4.

Given the complexity of factors (as shown in Figure 11.1) involved in any ethical consideration, there is not usually one right answer. Instead, it is important to find the most suitable outcome or compromise for all stakeholders involved.

Ethical matrices (as shown in Table 11.1 and 11.2) are frequently used to facilitate discussions and decisions. They ensure all concerns and relevant impact is captured and made transparent. As a first step, a matrix is populated with values (thus called value matrix). Then a second matrix (the consequence matrix) is developed to convey how certain actions impact on the identified values. These matrices also help to identify whether disagreements relate to different value priorities or understandings of the impact, which makes it somewhat easier to progress with a decision. Yet, no such decisions are ever easy.

Figure 11.1 The complex contexts of any ethical consideration. The figure illustrates the factors underpinning ethical points of view.

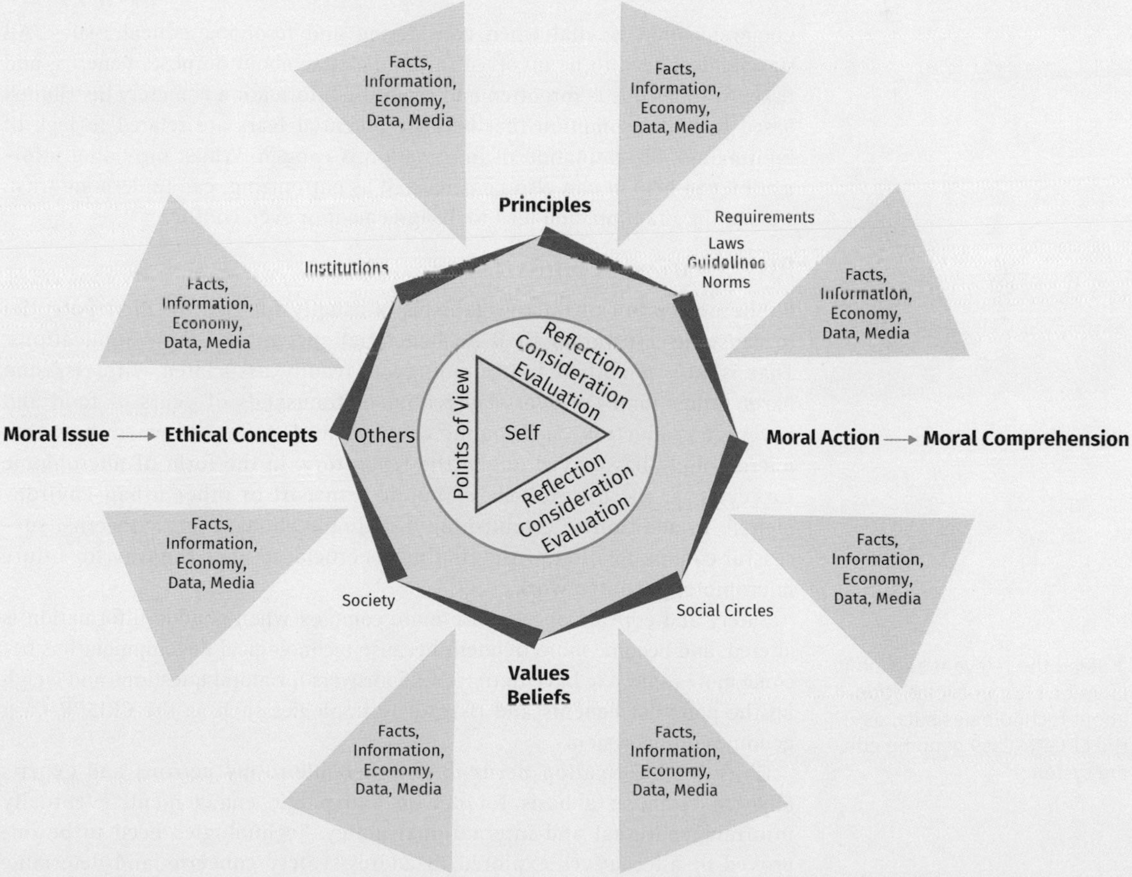

Table 11.1 The basic structure of an ethical matrix. A completed example is given for the context of patenting at Table 11.2.

Try to fill in the table as a value matrix and then a consequence matrix for some of the scenarios in this chapter or in the bigger picture panels in various chapters of the book. An example for issues related to patents in biotechnology is given at the end of this chapter. You should only look at it once you have attempted the task.

Stakeholder	Well-being	Autonomy/Dignity	Fairness/Justice
1			
2			
3			
4			
n			

11.1 Some contexts for ethical consideration

Communication is vital when considering and resolving ethical issues. All stakeholders need to be involved in the dialogue about purpose, benefits, and risks. At times this is forgotten and scientific information is merely distributed based on the assumption that because potential fears are related to lack of information, dissemination of information is enough. Whilst providing information can help, it can also be perceived as patronizing, can undermine trust, worsen the situation, and lead to disagreement or even conflict.

Risk awareness and safety

Public perception of microorganisms is usually focused on their potential to cause disease rather than on beneficial species and their applications. That is why microbial biotechnology is readily associated with risk and harm rather than the everyday reality of thousands of years of food and beverage production. Such public concern needs to be addressed, now that microbiology has moved out of the laboratory, in the form of microbiome surveys (e.g. the microbiome of public transport or other urban environments). In addition to establishing a dialogue about safety concerns, successful communication of the findings is crucial to pave the way for future microbiology-related work.

Safety and ethical concerns are more complex when genetic information is altered, and become more frequent because technological development has become more rapid. We have to try to find answers for moral questions and weigh up the potential benefits and risks of technologies such as the CRISPR-Cas9 genome editing system.

› Read the relevant section in Chapter 1 for an explanation about technologies such as the CRISPR-Cas9 genome editing system.

Objective information needs to be accessible to lay persons and experts alike to establish a basis for debate and public engagement, eventually informing national and international policy. Technologies need to be improved or alternatives explored to address safety concerns and determine future action.

Key points

- Bioethics issues are broad and complex in terms of safety, confidentiality, access to information, and intellectual property.
- Biotechnology has positive potential. Despite regulatory frameworks there are infringements and malicious intent.
- There is not usually one right answer in ethical considerations, but the most suitable outcome for all stakeholders needs to be found.
- Communication is vital and ethical matrices are frequently used to facilitate the discussion and decision.
- Public perception of microorganisms is usually focused on the potential to cause disease. Safety and ethical concerns are more complex when genetic information is altered.

Socioeconomics

Bioethical concerns are not only linked to the benefits and risks associated with technologies and their role, but also to the impact on society. While biotechnology has the potential to substantially contribute to solving global problems (e.g. food availability, fuel shortages) there is also a risk that poor people are affected negatively.

To minimize this risk, impact monitoring and assessment in a range of societies and societal contexts is required. Therefore policies and resources are put in place to strengthen biotechnology research in the public sector. The reality however is that much biotechnology research is conducted in the private sector and thus driven by markets. The **socioeconomic** context of countries where human problems (e.g. nutrition, healthcare) could be addressed with biotechnological solutions, differs from the context and problems of countries that debate the ethical frameworks. This disconnect means that any assessment of potential benefits or risks for poorer areas can only be extrapolated. So in reality, the poverty gap can actually increase and general risks may be unevenly distributed. An example of this is how to limit potential risks related to genetically improved organisms—which includes introducing potential allergens into food (see Case study 11.1). The necessary expertise and infrastructure for detection and analysis may not be accessible in poorer areas, or regulation can be. Even if regulations are implemented and adhered to, non-engineered food can be more expensive and therefore unaffordable.

Patenting

Patenting has become a focus and driver of the biotechnology sector, and adds to controversy and the complexity of ethical issues (see Table 11.2). In 1980 the US Supreme Court decided that genetically modified bacteria were patentable. Patents in Europe also protect microbiological processes or their products. This incentivized development and provided the basis for considerable growth of the entire biotechnology industry.

One of the controversial aspects is related to discriminating discovery and invention when it comes to genetic material. Should patents cover DNA? What about synthetic DNA? Patents related to CRISPR protect the basic technology as well some specific applications. The regulation is mixed for viruses: while the US Supreme Court ruled out patenting of isolated genetic products, Europe patents such isolates including viruses. Synthetic DNA does not occur in nature and is therefore patentable.

Such intellectual property rulings profoundly affect development of and access to molecular diagnostics. In general the extent and direction of research

Table 11.2 Ethical matrix for biotechnology patents (based on Forsberg et al. 2017). The matrix has been completed using the example of patenting.

Stakeholder	Well-being	Autonomy/Dignity	Fairness/Justice
Investors	Suitable business conditions	Influencing patenting issues/intellectual property	Fair return on investments made
Competitors	Suitable business conditions	Influencing patenting issues	Fair treatment of all companies
Commercial users of the patented invention	Access to high-quality output	Influencing patenting issues	Appropriate licensing
Consumers	Access to affordable output	Influencing patenting issues	Fair treatment
All human beings	No harm	Overall dignity and freedom	Fair treatment
Countries	Wealth creation	Democracy and sovereignty	Fair use of public resources
Future generations	Resource protection and precaution	No undue limitations	Fair use of benefits and share of burdens
Biosphere	Resource protection and precaution	No undue limitations	Fair use of ecosystem resources

and investment in research is impacted on and determined by patent regulation. Patents with narrow claims are more likely to allow enough room for invention, but patents with broad claims could stifle it, eventually limiting access to diagnostics, treatment, and medication. In any case the intellectual property landscape is unevenly distributed between developed and developing countries. This can only be resolved by considering and deciding much broader ethical questions.

Case study 11.1
What is on the menu?

The amount of endogenous allergens may be increased or new allergens may be introduced in foods constituted of or containing genetically modified microorganisms; engineered components as ingredients; traces from the production process (remember what we discussed in Chapters 3 and 4). This is a concern for the increasing number of allergy sufferers (2–8 percent of people) in developed and developing countries. Allergenicity risk assessments are required to inform consumers. Just eight food groups give rise to more than 90 percent of food allergens. Consumers will need to know what kind of allergens are contained in a product in order to make a purchase decision. Assessing allergenicity can be challenging, but it is generally agreed that the risk is low if the protein is not homologous to any known allergens, is contained only in low amounts, and is rapidly digested upon consumption. The Food and Agriculture Organization of the United Nations together with the World Health Organization has developed guidelines and recommendations (Codex Alimentarius) including a framework for safety assessment of foods produced using genetically modified microorganisms. Such assessments require resources (technology, trained analysts, infrastructure, paper trails, labelling, and monitoring) and therefore increase the costs of the food. As a consequence, the market becomes more limited and regulation is likely to be undermined.

How could infringements be monitored? Do you think consumers might be in a position to monitor their food for allergens at some point?

Scientific approach panel 11.1
Synthetic life

In 2010 **Craig Venter** and his team succeeded in creating a synthetic cell harbouring a computer-designed genome. This was an important milestone on the way to identifying the minimal cellular machinery necessary to sustain cellular life.

The cells (see Figure SA 11.1(a)) grow under the control of the synthetic genome, produce *Mycoplasma mycoides* proteins, and have its phenotype. At some point, due to natural turnover, all cell components will have been produced based on the information encoded on the synthetic genome.

Given that the cytoplasm and every part of the cell other than the genome, are natural and not synthetic, this synthetic cell is technically not synthetic life. Synthetic life would require that all components are synthesized separately and assembled to form a viable and growing cell. This puts everything in a somewhat different yet still complex ethical context.

The team reached a further milestone in 2016 by minimizing the genome of *Mycobacterium mycoides*, resulting in viable cells (see Figure SA11.1(b)) containing only 473 genes.

The successful yet controversial patenting of such synthetic organisms and their components raises the question, 'Who owns life?' What do you think? Why might someone disagree with you?

❯ See Chapter 6 for the description of how the synthetic genome of *Mycoplasma mycoides* can be transplanted into *Mycoplasma capricolum* cells, which subsequently lose their genomes.

Figure SA 11.1 (a) **Colony of *Mycobacterium mycoides JCVI-syn1.0* arising from a single cell harbouring a synthetic genome.** The blue colour is not the natural colony colour, but a chemical marker used for selection. (b) **Scanning electron micrograph of *Mycobacterium mycoides JCVI-syn3.0* cells.** Each cell contains 473 genes and divides every three hours.

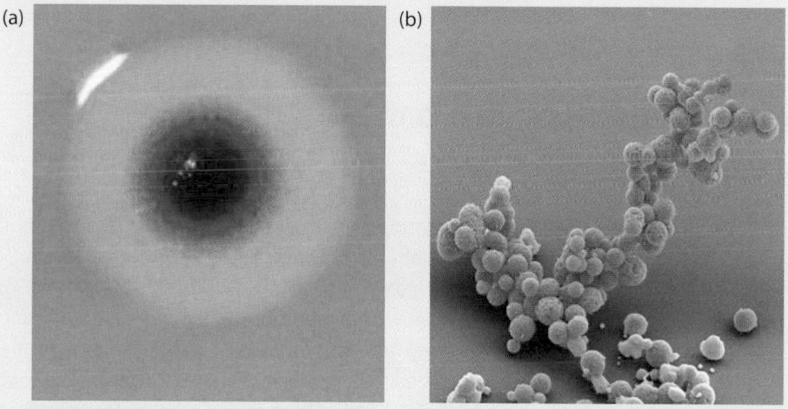

The need for considerable financial investment in protecting intellectual property is one factor contributing to uneven distribution. Managing such property well is crucial for institutions and countries. Protection is discontinued once the full term of a patent comes to an end, and then everyone can use previously protected processes to produce sought after products. The sector of expensive pharmaceutical products has been fast-growing globally and adequate patient access to medication is a challenge. One would think that biosimilars, cheaper versions of such medicines produced after a patent has expired, are the solution to fairer access by patients. This is not necessarily the case because market access is limited for a number of reasons. Regulatory systems and pricing structures can discriminate against imports. The production process in itself is still expensive and has to undergo the full approval process. Moreover, it is difficult to compete with the well established marketing and sales infrastructure of the initial innovator and with the reputation of the known brand of medication.

Diagnostic information

› In Chapter 7 we covered the advantages of sharing diagnoses swiftly via networked devices to allow correct treatment without delay for the benefit of the patient and potentially the community.

In addition to ethical issues around patient access to medication and to diagnostics, there are ethical issues around how diagnostic information is stored and used—matters which relate to privacy, confidentiality, and consent.

If this sensitive information is shared too broadly or without contextualizing information, considerable damage can be done to individuals, their family, and other environment. Health and life insurance may become impossible to obtain, employment may be terminated. Health information affects eligibility for a mortgage, having a home, and sustaining a relationship. Unsecured storage of biological information is more of a problem than the very rare incidents of hacking.

We cannot yet identify a specific human being based on their microbiome information, but there are already concerns regarding privacy. The fingertip microbiome can be linked to usage of certain keys of a computer keyboard. Inhabitants of a room can be tracked down based on the room's microbiome. The forensic use of microbiome data may therefore not be too far into

The bigger picture panel 11.1

Biosimilars

The reputation of an established pharmaceutical brand limits the use of biosimilars substantially. This may come as a surprise given that healthcare costs can be prohibitive and need minimizing. We know that approved biosimilars used in healthcare settings are safe and effective. For instance, the European Medicines Agency has issued a regulatory framework for registration of biosimilars. The framework requires proof of similarity to the already authorized original biological medicinal product based on a comprehensive comparability exercise. Biosimilars are thus not sourced from the unregulated black market. Yet, subconsciously it can feel like using a fake. There can be nagging doubt—can we trust a fake? After all, our health or that of a loved one is at stake. Would you want it injected? These rhetorical questions are just an example of how complex ethical issues are, and that we do not always approach them rationally. At least we have the opportunity to take a step back and where necessary access, consider, and analyse information to come to an informed decision. The situation is very different and worrying for those people who cannot access information or are unable to understand it, and have to trust what they are told. In this example in and in all other (microbial) biotechnology-related ones, we need to use our position to make a positive contribution to the debate and to establishing suitable ethical frameworks that benefit everyone.

What are the issues in this complex situation, and are there any that have not yet been alluded to? What are the challenges in addressing any of the issues?

the future. Third-party access to microbiome data therefore needs to be proactively restricted, and storage of the information must be safe in adherence with the regulatory framework regarding genetic information, which protects human data. However microbial DNA information is not human and usual storage arrangements put the individual at risk of identification. Security measures should be required for microbiome data to protect privacy. For the time being our rights to our microbiome are not protected.

Key points

- Bioethical concerns also relate to biotechnology's impact on society.
- Microbiological processes, products, and genetically modified bacteria as such are patentable, as are synthetic DNA and cDNA.
- Intellectual property rulings affect research and access to its benefits.
- When the term of a patent comes to an end everyone can use what was previously protected.
- There are ethical issues around patient access to diagnostics and how diagnostic information is stored and used regarding privacy, confidentiality, and consent.

11.2 Some unethical contexts

Even if the ideal ethical and regulatory frameworks were in place there will be problems because not everyone adheres to them, either from ignorance, or worse, due to criminal intent. Intellectual property rights are being infringed. Unregulated products of low quality are sold that can cause serious harm. Using information or technology to cause harm on purpose (e.g. during times of war) goes back many centuries.

Infringements

Any law or regulation can always be ignored, which in itself is unethical. Whether it is worth the risk for the perpetrator depends on the contexts. Illegally obtaining biological technology and undermining patents for the production of pharmaceuticals does not result in profit unless the pharmaceuticals are sold on unregulated black markets given legitimate market access is a challenge which we have already discussed for biosimilars. The sale of data (e.g. patient data) can generally be profitable in terms of the exploitation/application or limiting competitors' access to the data. Conducting experiments prohibited by law will not produce publishable scientific output leading to esteem—scrutiny within the scientific community would not allow it. However, scientific output may result in developing technology or products which could be profitable on unregulated markets.

Let us now consider in more detail the illegal acquisition of promising biological resources.

The 2014 Nagoya Protocol is an international framework which regulates the fair use of genetic resources. This includes microorganisms. Even though the framework is hardly applicable (given the concept of microbial biological diversity) it has the unintended potential to limit microbial research to the detriment of everyone. Trying to regulate access to microbial diversity is futile in light of the natural frequent **horizontal gene transfer**, which allows microbes to readily spread crucial features (e.g. entire metabolic pathways) in the microbial community.

Microorganisms contained in environmental samples have very limited monetary value. By contrast, characterized isolates can be valuable as they require a considerable financial investment. Isolating one bacterial strain can cost up to 10,000 Euros, and only about ten strains in one million isolates produce a natural compound of interest. Therefore, the cost benefit ratio is comparatively low.

Bioweapons and bioterrorism

Biological weapons are not a new concept. Nearly 2,000 years ago Greek soldiers intentionally contaminated the drinking water of their enemies by throwing dead animals into wells. In modern times specifically produced bioweapons are used (e.g. anthrax in the First World War and plague in the Second World War). In addition to biological warfare, there is also the danger to civilians in the context of bioterrorism where microorganisms or isolated toxins are released to cause mass destruction (e.g. anthrax attacks via the US postal system in 2001). Given that we also need food, bioweapons against livestock (e.g. brucellosis) and crops (e.g. *Fusarium* sp.) are a danger to human lives. Bioweapons are cheap, difficult to detect, and can be easily spread over large areas. Natural or engineered microorganisms of choice for bioweapons are highly contagious and result in high mortality. Bioterrorism attacks are a substantial drain on a health care system and can have a destabilizing effect on the economy. Society and everyday life may be seriously disrupted due to large-scale panic, even if only a few people fall ill or die. The enemy is invisible and does not stop at borders. Some countries have bioweapon programmes in place for research and defence, while others are suspected to possess bioweapons. Antibiotic-resistant and vaccine-subverting strains of known pathogens (e.g. smallpox, tularaemia) are a focus of bioweapon development. That is why training exeracises are conducted (such as seen in Figure 11.2) and protective equipment (such as in Figure 11.3) is stocked.

Figure 11.2 Individuals wearing protective suits during training for how to respond to a bioterrorist attack.

iStock.com/BenDC

Figure 11.3 Components of a protective suit worn in an Ebola outbreak.

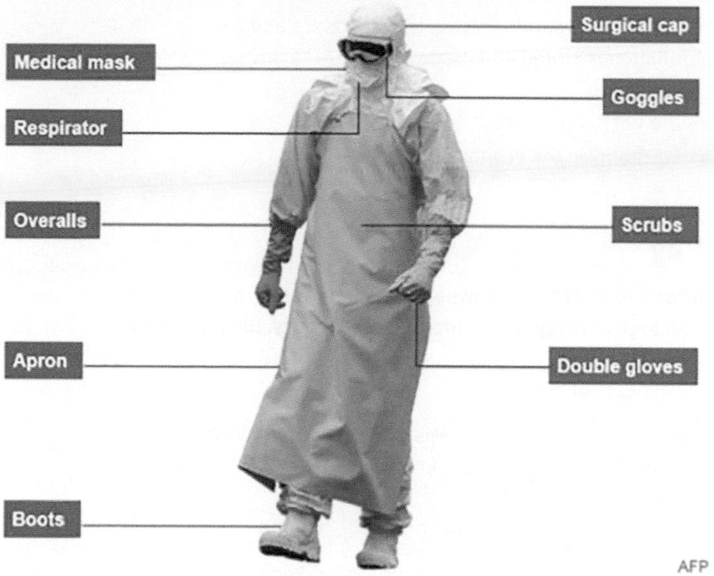

DOMINIQUE FAGET/AFP via Getty Images

Key points

- Any law or regulation can always be ignored, which in itself is unethical.
- Biological weapons are not a new concept but go back many centuries.
- In bioterrorism, highly contagious and virulent natural or engineered micro-organisms or isolated toxins are released to cause mass destruction.
- Bioterrorism attacks destabilize the economy and cause panic.

Chapter Summary

- Bioethics issues are broad and complex.
- Biotechnology has positive potential.
- There is usually not one right answer in ethical considerations, but the most suitable outcome for all stakeholders needs to be found.
- Communication is vital and ethical matrices are frequently used to facilitate the discussion and decision.
- Public perception of microorganisms is usually focused on the potential to cause disease. Safety and ethical concerns are more complex when genetic information is altered.
- Microbiological processes, products, genetically modified bacteria, synthetic DNA, and cDNA are patentable.

- When the term of a patent comes to an end everyone can use what was previously protected.
- There are ethical issues around patient access to diagnostics and how diagnostic information is stored and used with regard to privacy, confidentiality, and consent.
- Biological weapons are not a new concept.
- Bioterrorism attacks destabilize the economy and cause panic.

Further Reading

Dudley, J. P. and Woodford, M. H. (2002). 'Bioweapons, Biodiversity, and Ecocide: Potential Effects of Biological Weapons on Biological Diversity. BioScience 52: 583–92.
This paper reviews all aspects of bioweapons.

Forsberg, E-M. et al. (2017). 'Patent Ethics: The Misalignment of Views Between the Patent System and the Wider Society'. Sci Eng Ethics 24:5 1551–76; DOI 10.1007/s11948-017-9956-5
This paper discusses issues around patenting and public opinion.

Overmann, J. and Hartman Scholz, A. (2017). 'Microbiological Research Under the Nagoya Protocol: Facts and Fiction'. Trends in Microbiology 25: 85–8.
This paper explores the Nagoya Protocol in the context of microbial diversity.

Verbeken, G. et al. (2014). 'Call for a Dedicated European Legal Framework for Bacteriophage Therapy'. Arch. Immunol. Ther. Exp. 62: 117–29.
This paper exemplifies the issues around regulation.

Discussion Questions

11.1 Discuss why there is usually no single right answer to ethical considerations.

11.2 Discuss why public engagement with science is important.

11.3 Discuss why cooperation and regulatory frameworks across borders are essential.

GLOSSARY

A

Acetogenesis A process which produces acetate from carbon dioxide and an electron source.

Acetogens An organism which can produce acetate.

Agonist A compound binding to a receptor, resulting in a response.

Alloy A mix of at least two elements, one being a metal.

Analyte A substance which is being identified and measured.

Anamnesis A patient's account of their medical history.

Anion exchange A chromatography technique which separates proteins according to their charge. In anion exchange the resin used contains positively charged groups.

Anode A positive electrode.

Antagonist A compound interfering with the agonist.

Antibiotics Compound that is active against microbial life.

Aptamers Peptides or oligonucleotides which bind to a specific target molecule.

Average rate of cell division (chapter 2 page 15)

B

Bacteriophages Viruses which infect bacteria.

Bacteroids Structures within plant roots nodules which carry out nitrogen fixation.

Baffles Vertical strips of metal mounted against the wall of a reactor which help to reduce vortexing and swirling and aid aeration.

Batch culture (chapter 2 page 8)

Batch process A closed system of culture where medium is not added or removed.

Biofilm Community of microorganisms growing on surfaces, protected by secreted polymers.

Biologics Products obtained from natural sources and produced by biotechnological methods.

Bioreactor (chapter 2 page 8)

Bioreceptor Any compound which binds to another specific compound when part of a biosensor.

Bioremediation The development, use, and regulation of biological systems (i.e. microorganisms) to decontaminate environments.

Biosensors A biological recognition element.

Built environment Human-made environment.

C

Capillary action Liquid flows quickly into a narrow space.

Carboxysome A compartment present in photosynthetic cells which concentrates CO_2 within its interior, where it is incorporated into sugars.

Cathode Negative electrode.

Cation exchange A chromatography technique which separates proteins according to their charge. In cation exchange the resin used contains negatively charged groups.

Cell compartment Areas in the cytosol of a eukaryotic cell which are surrounded by a membrane.

Cell growth rate (chapter 2 page 15)

Chemoheterotrophic Using organic compounds as carbon and energy source.

Chemolithotrophic Using inorganic compounds as energy source.

Chemostat A culture system where new nutrients are added in the form of fresh medium at a specific rate, culture and spent medium is removed at the same rate.

Chemoorganotrophy Energy production, where an organic compound serves as the initial electron donor.

Codon usage The frequency of use of triplicate nucleotides which specify an amino acid residue.

Continuous culture (chapter 2 page 8)

Continuous process A method of production which is used to make products without interruption.

Contois model (chapter 2, page 20)

Craig Venter American biotechnologist, leading the first draft sequence of the human genome.

D

Deceleration phase The transition from exponential growth phase to stationary growth phase.

Diazotroph Microorganisms which are capable of nitrogen fixation.

Dilution rate (chapter 2 page 25)

DNA computing Computing based on DNA rather than silicon based.

Dongle Small computer hardware that fits into a port and adds functions.

Doubling time The length of time it takes for the cell population or biomass to double.

Downshift During growth, cells are moved from a good carbon source such as glucose to a poor carbon source such as cellulose.

E

Ecotoxic Chemically hazardous to the environment.

Electrochemical corrosion Destruction of metal when electrons from atoms at the surface of the metal are transferred to a suitable electron acceptor.

Electrolyte Compound that dissociates into ions, therefore transporting electric charge.

Endophytes Microbes which are able to penetrate the plant cell tissues and live inside the plant in locations such as the roots, leaves, and fruits, without causing any disease symptoms.

Endotoxin Lipid A present in the outer membrane of Gram-negative bacteria.

Enhancers A 50–1500 bp region of DNA that can be bound by proteins to increase the likelihood that transcription of a particular gene will occur.

Enzyme-linked immunosorbent assays A technique used to detect the presence of a substance using an antibody or antigen.

Erythrocyte A Red blood cell.

Exonuclease An enzyme which removes nucleotides from the end of a RNA or DNA molecule.

Exponential growth When cells growing in culture are doubling in population size over a specific time interval.

F

Fed-batch culture (chapter 2 page 8)

Fed-batch process A culture system where fresh media can be added, but spent medium is not removed.

Fluxes The rate of turnover of molecules through a metabolic pathway.

G

Generation time (doubling time) This is the length of time it takes for the cell population or biomass to double.

Glycosylation The modification of an organic molecule by the enzymatic addition of a sugar molecule.

Good Manufacturing Practice Minimum standards for everything related to the production process of pharmaceutical products in order to ensure consistent high quality.

GRAS Organisms used in biotechnology which are generally regarded as safe.

H

Hartig net Sheath of fungal hyphae surrounding the root cortical cell seen in ectomycorrhizal fungi.

Headspace The volume above the media in the reactor.

Heterocysts Specialized cells in cyanobacteria which are capable of nitrogen fixation.

Heterofermentative A fermentation which results in a number of end products.

Heterologous proteins Proteins which are expressed as a result of recombinant DNA technology.

Heteropolysaccharides A polysaccharide consisting of more than one type of sugar monomer.

High-throughput Simultaneous processing of a large number of samples.

Homofermentative A fermentation which results in a single end product.

Homologous recombination The exchange of nucleotide sequences between similar or identical molecules of DNA.

Homopolysaccharides A polysaccharide consisting of one type of sugar monomer.

Horizontal gene transfer Movement of genetic material to a cell that is not offspring.

Host species Organisms which are used to produce a desired product.

Hydration shell Water layer surrounding biomolecules.

Hydrolysis A chemical reaction which uses water to break chemical bonds.

Hydrophobic core Stabilizes folded state by not exposing hydrophobic surfaces.

Hydrothermal vents Openings in the seafloor where seawater meets magma.

Hyperglycaemic inhibitor Prevents high blood sugar.

Hyphae The filamentous structure of a eukaryotic fungus or prokaryotic actinobacterium.

I

Immobilization Cells which are physically trapped to a solid support.

Immunodetection The detection of molecules using antibodies.

Impellers A rotating structure used to move liquid within a reactor.

Inclusion bodies Insoluble protein aggregates.

Inducers A molecule which induces gene expression.

Inducible promoter Requires inducer for transcription to commence; expression not constitutive.

Inhibitors A molecule which inhibits gene expression or the activity of an enzyme.

Interaction hydrophobic chromatography A chromatography technique which separates proteins on the basis of their hydrophobicity.

Intrinsic flexibility Allowing conformational fluctuations.

K

Klenow fragment Large fragment of DNA polymerase I.

L

Lignocellulose A complex macromolecule containing cellulose, hemicellulose, and lignin.

M

Malt Germinated cereal grain such as barley.

Mesophilic Organisms which grow between 25–45°C.

Metabolic downshift Reduced transcription and translation.

Metabolic flux analysis Modelling of product formation using the biochemical pathways of the cell.

Metabolite A substance produced or modified by a metabolic process.

Metagenomic Relates to genetic material in the environment.

Methanogenesis The production of methane by microbes.

Mismatch repair A DNA repair system which both recognizes and repairs misincorporated nucleotides during replication and recombination.

Monoclonal antibodies Identical antibodies produced by a single cell line.

Mordant Compound that combines with a dye to fix it in a material.

Mortality Death.

Multidrug efflux pumps Transporter protein complexes that flush out compounds.

Mutagen A chemical which can cause DNA damage.

Mycorrhizosphere A unique microbial community present in the soil adjacent to the fungal hyphae.

N

Nanomaterials Materials which consist of single units between 1 and 1,000 nanometres in size.

Nitrifier Uses ammonia or NO_2 as electron donor.

Nitrogenase An enzyme present in some prokaryotes which can reduce dinitrogen gas to ammonia.

O

Oleaginous Organisms which can produce oil.

Oocyte Egg cell.

Origin of replication A specific sequence of DNA where replication of the DNA molecule is initiated.

P

Patent The grant of a property right by the government to an inventor for a set period of time.

Periplasmic extraction The removal of substances from the periplasm of Gram-negative bacteria.

Phage display A technique used to study protein interactions, where one protein partner is displayed by a bacteriophage.

Phenotype The observable characteristics of an organism due to the expression of genes.

Photobioreactors Reactors used for the cultivation of phototrophic microbes such as microalgae and cyanobacteria.

Photolithoautotrophic Using light as energy source and carbon dioxide as carbon source.

Phytosterols Plant sterols.

Plantibodies A recombinant antibody which is produced by plants as hosts.

Plasmids Extrachromosomal DNA capable of autonomous replication.

Plasmid segregation Divides up plasmids for daughter cells.

Post-translational modifications The modification of proteins after they have been translated.

Precursors Compounds that are transformed to become the final product of interest.

Primary metabolites Metabolites which are produced during exponential phase of growth.

Probiotic A microbe which can confer a health benefit when taken as a food or supplement.

Productivity The amount of product produced during a process over a specified period of time.

Protected-origin foods Foods produced in specific geographical locations which have protected names.

Pulsed polarization Change of polarity in intervals.

Q

Quorum sensing Regulation of gene expression based on cell-population density.

R

Random mutagenesis A process of mutagenesis inserts mutations randomly.

Reactor A vessel used for the culturing of cells.

Reader device Apparatus to obtain a measurement output.

Recombinant antibodies Antibodies which are made through recombinant DNA technology.

Recombinant vector A plasmid (typically) which has artificially inserted foreign DNA.

Reporter gene A gene which is attached to the regulatory sequence of another gene of interest to study gene expression.

Rhizosphere The area of soil directly adjacent to the plants roots.

Rhizosphere effect The increased population of microbial cells present in the rhizosphere compared with the bulk soil.

S

Secondary metabolites Metabolites which are produced during stationary phase of growth.

Secretion system Protein complexes present in cell membranes through which substances can pass.

Seeps The flow of a liquid through holes or pores in material.

Selectable marker A gene which is used to select for a plasmid inside a host cell.

Serological analysis A diagnostic test which identifies the presence of antibodies or antigens.

Sigma factor Subunit of RNA polymerase, specific to promoter binding.

Signal sequence A short peptide which directs a newly formed protein to a secretory pathway.

Silent clusters Groups of genes that are not expressed.

Single cell protein The whole biomass of a microorganism grown for the production of food.

Single nucleotide polymorphisms Genetic variation based on one nucleotide.

Site-directed mutagenesis A Process of mutagenesis which targets a specific site of DNA.

Socioeconomic Interaction of social and economic factors.

Solid state Where cells are grown in solid media without the further addition of water.

Steady state As culture is removed from a reactor and nutrients are added, the biomass in the fermenter vessel remains constant.

Strain Subspecies variant.

Submerged culture Where cells are grown in liquid media.

Substrate-level phosphorylation Phosphate group is transferred from substrate to ADP.

Sulphur and iron oxidizer Uses H_2S, S or Fe^{2+} as electron donor.

Surfactants Compounds that lower the surface tension.

T

Thermophilic Organisms which grow between 50–60°C.

3D bioprinting Three dimensional printing of biological materials.

Transducer element Converts a biological response into an electrical, optical, or thermal signal.

Transposons Mobile genetic elements.

Two-component regulator (Chapter 8, section 8.4 page 26)

U

Ultrafiltration A process of filtration which uses a semi-permeable membrane.

Unculturable Organisms which cannot be grown in laboratory culture.

Upshift During growth cells are moved from a poor carbon source to a good carbon source.

W

Wort Liquid made from partially germinated barley grains used to make beer.

Y

Yield The relationship between product formed and substrate consumed.

Z

Zwitterion Molecule with separate positively and negatively charged groups.

BIBLIOGRAPHY

Chapter 1

Dahod, S. K., et al. (2010). 'Raw Materials Selection and Medium Development for Industrial Fermentation Processes'. In: *Manual of Industrial Microbiology and Biotechnology*, ASM Press, ISBN: 9781555815127.

Doran P. M. (2012). *Bioprocess Engineering Principles*. 2nd edn, Academic Press, ISBN-10: 012220851X.

Niazi, S. K. and Brown, J. L. (2017). *Fundamentals of Modern Bioprocessing*. CRC Press, ISBN: 9781138893290.

Okafor, N. (2017). 'Fermenter and fermenter operation'. In: *Modern Industrial Microbiology and Biotechnology*, CRC Press, ISBN: 9781138550186.

Rehm, H.-J and Reed, G. (2008). *Biotechnology: Bioprocessing*, Vol 3, 2nd edn, VCH Verlagsgesellschaft GmbH, ISBN: 9783527283132; DOI: 10.1002/9783527620845.

Spier, M., et al. (2011). 'Application of different types of bioreactors in bioprocesses'. In: *Biorectors, Design, Properties and Applications*, Nova Science Publishers, Inc. ISBN: 9781621001645.

Steel, B. and Stowers, M. D. (1991). 'Techniques for selection of industrially important microorganisms'. Annual Reviews in Microbiology 45: 89–106.

Volmer, J., Schmid, A., and Buehler, B. (2015). 'Guiding bioprocess design by microbial ecology'. Current Opinion in Microbiology 25: 25–32.

Waites, M. J., et al. (2001). *Industrial Microbiology: An Introduction*. Blackwell Science Ltd, ISBN: 0-632-05307-0.

Weuster-Botz, D. (2000). 'Experimental design for fermentation media development: statistical design or global random search?' Journal of Bioscience and Bioengineering 90: 473–83.

Chapter 2

Waites, M. J., et al. (2001). *Industrial Microbiology: An Introduction*. Blackwell Science Ltd, ISBN 0-632-05307-0.

Chapter 3

Adrio, J. L. and Demain, A. L. (2014). 'Microbial enzymes: tools for biotechnological processes'. Biomolecules 4: 117–39.

Cheon, S., et al. (2016). 'Recent trends in metabolic engineering of microorganisms for the production of advanced biofuels'. Current Opinion in Chemical Biology 35: 10–21.

Doran, P. M. (2012). *Bioprocess Engineering Principles*. 2nd edn, Academic Press, ISBN-10: 012220851X.

Gurung, N., et al. (2013). 'A broader view: microbial enzymes and their relevance in industries, medicine, and beyond'. BioMed Research International, Article ID 329121 http://dx.doi.org/10.1155/2013/329121.

Kang, A. and Lee, T. S. (2015). 'Converting sugars to biofuels: ethanol and beyond'. Bioengineering 2: 184–203.

Martien, J. I. and Amador-Noguez, D. (2017). 'Recent applications of metabolomics to advance microbial biofuel production'. Current Opinion in Biotechnology 43: 118–26.

Rehm, B. H. A. ed (2009). *Microbial Production of Biopolymers and Polymer Precursors: Applications and Perspectives*. Caister Academic Press, ISBN: 9781904455363.

Sanchez-Garcia L., et al. (2016). 'Recombinant pharmaceuticals from microbial cells: a 2015 update'. Microbial Cell Factories 15: 33; DOI 10.1186/s12934-016-0437-3.

Singh, R., et al. (2016). 'Microbial enzymes: industrial progress in 21st century'. Biotech 6: 174; DOI 10.1007/s13205-016-0485-8.

Tharali, A. D, Sain, N., and Jabez Osborne, W. (2016). 'Microbial fuel cells in bioelectricity production. Frontiers in Life Science 9: 252–66 http://dx.doi.org/10.1080/21553769.2016.1230787.

Wang, M., Jiang, S., and Wang, Y. (2016). 'Recent advances in the production of recombinant subunit vaccines in *Pichia pastoris*'. Bioengineered 7: 155–65; DOI:10.1080/21655979.2016.1191707.

Wen, E. P., Ellis, R., and Pujar, N. S. eds (2015). *Vaccine Development and Manufacturing*. John Wiley & Sons Inc. ISBN: 9780470261941; DOI: 10.1002/9781118870914.

zu Berstenhorst, S. M., Hohmann, H.-P., and Stahmann, K.-P. (2009). 'Vitamins and vitamin-like compounds: microbial production'. *Encyclopedia of Microbiology* 549–61; DOI 10.1016/B978-012373944-5.00161-9.

Chapter 4

Bagchi, D., et al. (2010). *Biotechnology in Functional Foods and Nutraceuticals*. CRC Press, ISBN: 9781420087116.

Hutkins, R. W. (2006). *Microbiology and Technology of Fermented Foods*. Wiley-Blackwell, ISBN: 9780813800189.

Joshi, V. K and Singh, R. S. (2012), *Food Biotechnology*. I K International Publishing House, ISBN-10: 9381141495.

Kumar, R., Vikramachakravarthi, D., and Pal, P. (2014). 'Production and purification of glutamic acid: a critical review towards process intensification'. Chemical Engineering and Processing 81: 59–71.

Lee, B. H. (2014). *Fundamentals of Food Biotechnology*. 2nd edn, Wiley-Blackwell, ISBN: 9781118384916.

Ray, R. C. and Montet, D. (2014). *Microorganisms and Fermentation of Traditional Food*. CRC Press, ISBN: 9781482223088.

Suman, G., et al. (2015). 'Single cell protein production: A Review'. International Journal of Current Microbiology and Applied Sciences 4: 251–62.

Wendisch, V. F. (2014). 'Microbial production of amino acids and derived chemicals: synthetic biology approaches to strain development'. Current Opinion in Biotechnology 30: 51–8.

Chapter 5

Lea-Smith, D. J., et al. (2015). 'Major contribution of cyanobacterial alkane production to the ocean hydrocarbon cycle'. Proceedings of the National Academy of Sciences 112: 13591–6.

Lea-Smith, D. J., et al. (2016). 'Hydrocarbons are essential for optimal cell size, division and growth of cyanobacteria'. Plant Physiology 172: 1928–40.

Saar, K. L., et al. (2018). 'Enhancing power density by biophotovoltaics by decoupling storage and power delivery'. Nature Energy 3: 75–81.

Chapter 6

Baers, L. L., et al. (2019). 'Cyanobacterial proteome mapping reveals distinct compartment organisation and metabolism dispersed throughout the cell'. Plant Physiology 181: 1721–38.

Vasudevan, R., et al. (2019). 'CyanoGate: a modular cloning suite for engineering cyanobacteria based on the plant MoClo syntax'. Plant Physiology 180: 39–55.

Chapter 7

Capelli, L., Sironi, S., and Del Rosso, R. (2014). 'Electronic noses for environmental monitoring applications'. Sensors 14: 19979–20007.

Dolen, V. (2017). 'Changing diagnostic paradigms for microbiology. Report on an American Academy of Microbiology Colloquium held in Washington, DC, from 17 to 18 October 2016'. American Academy of Microbiology, Washington DC.

Gui Q., et al. (2017) 'The application of whole cell-based biosensors for use in environmental analysis and in medical diagnostics'. Sensors 17: 1623.

Niemz, A., Ferguson, T. M., and Boyle, D. S. (2011). 'Point-of-care nucleic acid testing for infectious diseases'. Trends in Biotechnology 29: 240–50.

Ravikumar, S., et al. (2017). 'Engineered microbial biosensors based on bacterial two-component systems as synthetic biotechnology platforms in bioremediation and biorefinery'. Microb Cell Fact 16: 62.

Wang, Y., et al. (2017). 'Application of nanodiagnostics in point-of-care tests for infectious diseases'. International Journal of Nanomedicine 12:4789–803.

Chapter 8

Barea, J. M. (2015). 'Future challenges and perspectives for applying microbial biotechnology in sustainable agriculture based on a better understanding of plant-microbe interactions'. Journal of Soil Science and Plant Nutrition 15: 261–82.

Berg, G. (2009). 'Plant–microbe interactions promoting plant growth and health: perspectives for controlled use of microorganisms in agriculture'. Applied Microbial Biotechnology 84: 11–18.

Bisen, P. S., et al. (2012). 'Microbes in Agriculture'. In: *Microbes: Concepts and Applications*, John Wiley & Sons Inc. https://doi.org/10.1002/9781118311912.

Okafor, N. (2017). 'Manufacture of *Rhizobium* Inoculants'. In: *Modern Industrial Microbiology and Biotechnology*, CRC Press, ISBN: 9781138550186.

Toyota, K. and Watanable, T. (2013) 'Recent trends in microbial inoculants in agriculture'. Microbes and Environments 4: 403–4.

Chapter 9

Anitori, R. P. ed (2012). *Extremophiles: Microbiology and Biotechnology*. Caister Academic Press, ISBN: 9781904455981.

Durvasula, R. D. and Subba Rao, D. V. (2018). *Extremophiles: From Biology to Biotechnology.* CRC Press, ISBN 9781498774925.

Krüger, A., et al. (2018). 'Towards a sustainable biobased industry—highlighting the impact of extremophiles'. New Biotechnology 40: 144–53.

Lee, N. ed (2018). *Biotechnological Applications of Extremophilic Microorganisms.* De Gruyter, ISBN: 9783110427738.

Raddadi, N., et al. (2015). 'Biotechnological applications of extremophiles, extremozymes and extremolytes'. Appl Microbiol Biotechnol 9: 7907.

Rampelotto, P. H. ed (2016). *Biotechnology of Extremophiles: Advances and Challenges.* Springer, ISBN: 9783319135205.

Chapter 10

Loto, C. A. (2017). 'Microbiological corrosion: mechanism, control and impact—a review'. Int J Adv Manuf Technol 92: 4241–52.

Sterflinger, K. and Piñar, G. (2013). 'Microbial deterioration of cultural heritage and works of art—tilting at windmills?' Appl Microbiol Biotechnol 97: 9637–46.

Volodymyr, I. and Stabnikov, V. (2017). *Construction Biotechnology Biogeochemistry, Microbiology and Biotechnology of Construction Materials and Processes.* Springer, ISBN: 9789811014444.

Chapter 11

Adenle, A. A., Morris, E. L., and Murphy, D. J. eds (2017). *Genetically Modified Organisms in Developing Countries, Risk Analysis and Governance.* Cambridge University Press, ISBN: 9781107151918.

Ludlow, K., Smyth, S. J., and Falck-Zepeda, J. eds (2014). *Socio-Economic Considerations in Biotechnology Regulation.* Springer, ISBN: 9781493943852.

Stevens, H. (2016). *Biotechnology and Society, An Introduction.* University of Chicago Press, ISBN: 9780226046013.

Storz, U., Flasche, W., and Driehaus J. (2012). *Intellectual Property Issues, Therapeutics, Vaccines and Molecular Diagnostics.* Springer, ISBN: 9783642295256.

INDEX

Note: Tables and figures are indicated by an italic *t* and *f* following the page number.

C

E

F